Cracking the Darwin Code

Exploring the Non-Scientific Foundations of Deep-Time and Evolution

Andrew M. Sibley

Fastnet Publications

First published in Great Britain in 2013

Fastnet Publications
Mill Close, Mounthill
Colyton, Devon. EX24 6EU

ISBN: 978-0-9562146-1-4

Printed by Lightning Source

By the same author:
- *Zion's New Name*, Colyton: Fastnet Publications, 2009
- *Restoring the Ethics of Creation*, Camberley: Anno-Mundi Books, 2006

Cover images are in the public domain - for further information about cover images see page vii.

Contents

Foreword

By John D Matthews, PhD
Retired chartered geologist and Anglican LLM

Andrew Sibley's book answers a question that has bothered me for a long time. We cannot get away from Charles Darwin; we see his image when we shop using a £10-note; his theory of evolution seems to be pulled out to explain everything concerned with how our bodies work and where we came from. Any dissenting voices are fiercely silenced - as I know from personal, painful experience. But why is Darwinism and acceptance of deep time so important? The problem is that Darwin and like-minded colleagues had their own agenda. Darwin and friends were not the unbiased scientists looking for truth that many think they were. They were out to undermine Christian beliefs and doctrines by replacing them with other beliefs. Yet with this fixation there are brief moments when Darwin admits what he is doing. On page 2 of his book on *The Origins of Species by means of natural selection or the preservation of favoured races in the struggle for life*, he says that there is scarcely a fact (of biology and geology) that he will discuss that cannot be explained by some other route than natural selection. That other route is by direct creation by God and a recent world-wide Flood.

Andrew explains that the dismissal of a recent global Flood, the introduction of evolution into our education system and arguments that the earth is very old are not based on a rational assessment of facts. They are based on ideas and beliefs already present in the Ancient Middle East and Indian sub-continent, but brought into the modern Western world by Enlightenment philosophers. These were a closely-knit group of people with a shared objective to attack Christianity. Their attacks were often subtle, as Darwin admits, and they concealed their methods to avoid an honest exchange of ideas.

Andrew looks in detail at the influences upon Darwin that led him to promote a distortion of truth. Darwin had been heavily influenced by people like Charles Lyell who had developed a way of weaning readers off everything that Moses wrote about concerning how and why and where we came from. And Lyell did this by distracting his readers, just as Darwin did, from other facts and interpretations that lead to opposite conclusions. Both Darwin and Lyell were seemingly following Voltaire's idea of pursuing a slow and silent attack upon Christianity, as Darwin himself admits in a frank letter. So Darwin's code of attitude and conduct is shown for what it really is – borne out of a desire to avoid the claims of Christianity, and replace it with other beliefs that may be traced back all through the history of human existence. Darwin's code is cracked.

But we cannot leave it there. The challenge is to promote truth – that we are not evolved creatures, we are directly made by God.

Acknowledgements

Thanks to John Matthews for offering comments and advice and writing a Foreword, and Phillip Bell of Creation Ministries International (CMI) for also reading through a draft copy.

Malcolm Bowden of CSM has also written along similar lines in his book *The Rise of the Evolutionary Fraud*, and also Henry Morris' *The Long War against God*. I want also to note that a number of members of the Edinburgh Creation Group have been studying similar lines of thought to the work presented in this book. See for instance: Phil Holden *The Modern Rise of Paganism* and Paul James-Griffiths *The Roots of Evolution*. Both of these presentations sourced via http://edinburghcreationgroup.org.

Several chapters in this book are edited and adapted from work published elsewhere. Chapter 3 is edited, adapted and expanded from; Sibley, A., Creationism and millennialism among the Church Fathers', *Journal of Creation*, 26(3), CMI, pp. 95-100, December 2012. Chapter 13 is rewritten and adapted from Sibley, A, 'Bathybius and a Reign of Terror,' *Journal of Creation*, 23(1), CMI, pp. 123–127, April 2009. Chapter 9 is loosely based upon Sibley, A, 'FitzRoy, Captain of the Beagle, Fierce Critic of Darwinism,' *Impact*, 389, ICR, November 2005.

Front cover: Adapted image of Dagon taken from a bas-relief at the Musée du Louvre, Paris (*public domain*). Dagon was an East Semitic god, worshipped for instance by the Philistines, also by the Assyrians and Babylonians (*1 Samuel 5:2-7*). Samuel records how the idol fell from its place, with head and arms broken, when the Ark of the Covenant was placed in the pagan temple. Similarly, Tiamat was also part of the Ancient Middle Eastern pantheon as the semi-aquatic oceanic goddess of chaos. The image of Charles Darwin is a watercolour by George Richmond, late 1930s (*public domain*). Erasmus Darwin placed three scallop shells on the Darwin family crest with the words *E conchis omnia* (everything from shells). According to Hesiod the Greek god Aphrodite (Roman Venus) was carried ashore on a scallop shell.

Preface

The work presented in this book grew out of a desire to get to the root of the concept of evolution, together with its supporting pillar of 'deep time,' which covers millions of years in which gradual evolutionary change is supposed to play its undirected, creative role. The purpose for this research, carried out over many years, then is to seek to unlock the foundations of evolutionary change, and so perhaps understand why the ideas are so important to the modern secular mindset. The questions I seek to raise and explore here are whether acceptance of long geological ages arose because of the scientific data, or because of a prior metaphysical commitment, one that relates to a naturalistic worldview that is essentially religious in character. All societies and cultures have their foundational creation stories, and Darwinian evolution may be seen in that light as well. Such an account I believe derives from a naturalistic prior commitment, and one where the concept of nature is sufficiently ambiguous to support a polytheistic, pantheistic or an atheistic worldview.

The purpose of this book then is to trace the origin of evolution and long geological ages through history and to explore how much influence such pagan or *nature* beliefs have really played in the development of modern naturalistic science. Some of the ideas here may not be easy to grasp immediately and may require careful study, but I will try and give a basic overview in the introduction. Mention of fellow believers in this book who accept old earth creationism or theistic evolution is offered in a spirit of respectful dialogue.

1.

Introduction - Setting the Scene

Aristotle was a proponent of a deductive form of logic and a rationalist, to the extent that even God must have created out of necessity as an unmoved mover. Unfortunately, he also supported scientific ideas that we would not accept today such as geocentrism and spontaneous generation; regrettably ideas embraced by a number of the Church Fathers and later Christian thinkers. This book will touch upon the way in which Aristotelian geocentrism gave way to heliocentrism. But more significantly, it will examine the manner in which belief in the spontaneous generation of fossils gradually collapsed. This collapse was largely due to the commitment of a number of Christian scientists to both Scripture and geological evidence. Later, however, a form of spontaneous generation reappeared as a quasi-pagan belief in evolution through for instance the work of Erasmus Darwin, and is still present in the naturalistic belief that life arose from a primeval pond of organic soup.

Aristotle's deductive reasoning has also given way to induction and abduction; all of our knowledge claims must begin with prior commitments that are not supported directly by evidence, but begin in faith. Aristotelian deductive reasoning then ultimately fails, as do the arguments for geocentrism and spontaneous generation.

On the other hand Plato was a dualist and this sets up another set of problems for science, although these problems are not so well known. A dualism exists for instance in the separation of the spiritual and material into separate realms of thought. With influence from Plato, the Gnostics saw the material world as evil and the spiritual world as good. The focus of the mind was then to escape the present world of suffering, which is why incidentally the Athenians struggled to accept Paul's preaching regarding the resurrection of the dead *upon the earth* (see *Acts 17*). Plato saw the material world as merely an outward form as a shadow of a greater spiritual reality. The spiritual was considered more real and more important than the physical. The greatest and most noble calling for an elite few was to contemplate this spiritual realm. The Platonist' approach to science then may be one of disinterest, and this arguably hindered the development of science for many centuries. But Platonism shows itself in the concept of a strong separation between science and faith, through insistence upon methodological naturalism, ironically in this concept classical Gnosticism is inverted. Atheists such as Richard Dawkins seem to view material science as good, and the road to enlightenment, and the spiritual as

1

inherently evil. But the influence of Plato in this separation can be seen, and it does not fit well with a Christian understanding of the world.

Plato also used progressive dialogue to develop his ideas and this risks leading to relativism in science because concepts such as truth and justice may change their meaning. So, in two ways Platonism may undermine science, firstly because it may lead to disinterest in the material world, and secondly because the inherent dualism risks leading to relativism and post-modernism where the idea of truth is relative. Rather ironically, post-modernism may be viewed as the illegitimate child of scientific modernism because the latter undermines a divine basis for objective truth. Truth instead becomes subject to human whims. But science must hold to objective truth as a foundational commitment if it is to retain coherence. Christian doctrines about God have historically provided this solid foundation; i.e. mankind created in the image of God.

So one aspect of this book is to explore what might be termed the *problem of Plato* in science. Although it does not just stem from Plato, similar ideas are also present in other Greek thinkers such as Hesiod with his *Theogeny* and importantly other areas of Eastern mysticism such as Baal worship and Hinduism. The *Theogeny* (generation of the gods) was a poem that expressed how pagan gods arose through self-generation and reproduction; then extending down to life on earth.

As an example to better understand the problem of relativity in Plato's thinking consider the concept of nature. Nature as a concept is ambiguous and can mean different things; on the one hand nature is considered to be all there is; but on the other hand nature is ascribed god-like creative properties as an impersonal force or *power of generation*, for instance represented by *Mother Nature*, *Gaia* or other pagan or eastern beliefs. For atheists, the truth of science is also assumed to arise out of nature, and nature is held to be self-existing, but these are properties that ought to belong to God and they cannot be properly grounded in nature.[1] This ambiguity in nature, between atheism and something akin to idolatrous polytheism or pantheism, has also often appeared in the writing of Enlightenment philosophers such as Baruch Spinoza, but also alluded to quite strongly in the work of David Hume and Erasmus Darwin for instance. The pagan concept of a *power of generation* is really a precursor to the Darwinian evolutionary narrative and arises out of a pantheistic understanding of nature. But the concept of nature is ambiguous and potentially undermines science because truth becomes relative. So the contention here is that the narrative of a grand evolutionary progression of life is, in effect, linked to a pantheistic understanding of nature.

---

[1] See Clouser, R., *The Myth of Religious Neutrality*, Notre Dame; University of Notre Dame Press, 2005

Introduction

What may be shown is that in the modern period, with the rise of science, the works of Plato, and other Greek thinkers (and eastern mystical beliefs), have influenced the development and acceptance of evolution and provided the basis for acceptance of billions of years of history. For instance, the age of the earth, currently accepted by secular science as several billions of years old, correlates very closely with one day of Brahma. Is this a coincidence? It is possible to show that this is more than a coincidence because Enlightenment writers produced books that argued for this outcome, and these influential philosophers had a strong interest in pagan beliefs. The evidence suggests then that pagan and Hindu religious texts really have influenced the tenets of modern evolutionary science. There is a degree of irony here, in that while the literal interpretation of the Genesis account of origins has been deliberately excluded from having any place in origin science, on the basis that no religious texts can or should have anything to say about natural science, non-Christian religious texts have had a great deal of influence in determining the direction of some modern scientific claims.

In the second chapter, in order to set the scene, it has been necessary to examine ancient cultures, from before the time of Christ. The purpose being to considered the passage of time, evolutionary change and the nature of the universe. Research of this nature doesn't fit into neat stereotypes because Plato and Aristotle even influenced many of the Church Fathers. Indeed Plato has often been considered very close to Christian thinking, although we may note with St Paul that the devil may even appear as an angel of light (2 *Corn. 11:14*). We do though need to move beyond superficial appearances and look deeper into these questions. Following a brief look at the Church Fathers in Chapter 3 consideration is then given to important questions in the development of modern science up to the seventeenth century, before looking in detail at developments in the eighteenth and nineteenth centuries. It will be shown how influence from pagan texts, liberal politics, revolutionary considerations, and deistic faiths such as Unitarianism and Freemasonry, played their part in developing acceptance of deep time and evolution. Use of clever rhetoric, secret plans, false evidence, and bullying appear to have played their part.

The conclusions that will be presented here will be based on an interpretive framework that there is really a struggle between two opposing religious traditions that have a claim over science and society. Augustine saw this struggle as one between the City of God, (in *De Civitas Dei*) and the City of the World.[2] Henry Morris saw it as a long war against God, but

---

[2] See Plantinga, M. 'Methodological Naturalism?' in Pennock, R.T. (ed.) *Intelligent Design Creationism and its Critics: Philosophical, Theological and scientific perspectives*, Cambridge Massachusetts, The MIT Press, 2001, p. 339

it is really the same struggle as that found in the opening passage of Genesis. That is, between God and the serpent for the soul of mankind and control over creation, and also in Revelation where the dragon is seen pursuing the woman and her child resulting in a war in heaven (*Rev. 12*). In Revelation a terrifying beast is seen rising out of the sea opening his mouth in blasphemy against God, while a prostitute is seen riding the beast with the title of 'Mystery Babylon' on her forehead; later described as a 'city' that rules over the kings of the earth (*Rev 13, & 17:18*). The philosophy, science, and social policies of ancient pagan religions became encapsulated in the works of Greek philosophers, especially Plato with the *Timaeus*, and also in the *Republic* where Plato discussed the policies of a repressive city state *Polis*. Amongst all the classical writers from antiquity, the works of Plato seems to have gained the ascendancy in modern western culture with a very strong view of how an ideal city-state should be ruled, but it is an elitist state that does not value the quality of education for all.

There are also questions relating to *nature* and the origin of the Enlightenment. We may recall Paul's warning that in the last days people will have a form of godliness, but deny the power of God (*2 Tim. 3:1-5*). The concept of a religious path to enlightenment is found in many eastern religions as well as Greek pagan beliefs, and it has become a central theme of western philosophy in the last few hundred years. Peter also warned that scoffers will in the last days arise who will seek to deny the Creation and Flood to put out of their minds the fear of God and judgment (*2 Pet. 3: 1-7*). The concept of enlightenment is of central importance to the development of the secular, humanistic society that is so prevalent today, and this covers many aspects of thought. In the question of origins there are two mutually supporting pillars that are evolutionary theory and acceptance of long ages. Many of the ancient pagan beliefs in fact had a cyclical view of history where events played out according to repetitive, perhaps eternal cycles. This is in stark contrast to the short, linear passage of time, and a material creation out of nothing, that is presented in the biblical account. The neo-pagan influence on naturalistic science however is usually not discussed in the public sphere, especially its influence on popular scientific thinking.[3] But instead the Bible has faced much greater public scrutiny and attack. This book will attempt to redress the balance by exposing secular science to a careful examination of its origins.

---

[3] Although it is a far more common theme in academic circles, as can be seen by the number of books on Plato in university libraries.

2

Deep Time and Evolution in Antiquity

Deep Time in Antiquity

This chapter will consider the origin of belief in deep time in antiquity, particularly from Baal worship and Hinduism, and see how it compares to the Genesis account. Later chapters will show how a reawakening of these ancient beliefs in the eighteenth century fed into the development of geology as a science. In other words, ancient pagan mysticism informed belief in millions of years of geological change before anyone had even attempted to justify it scientifically.

Some of the oldest written documents from the ancient Middle East, apart from the biblical text, extended the timeframe for creation substantially beyond that given in the biblical account. The Sumerian King List (SKL) for instance has notable similarity with Genesis. It has eight kings ruling from before the Flood, but with an apparently greatly extended timeframe in contrast with the Mosaic Flood account. The eight pre-Flood kings were said to have ruled for 241,200 years.[1] Although the Genesis account has the age of the Patriarchs extending for hundreds of years, with Adam living for 930 years and Methuselah for 969 years, many of the ancient pagan texts seemed to have increased this substantially. However, if one compares the periods given in the SKL with the Genesis account, and taking into consideration the complexity of the Sumerian numbering system, it may be possible to find a degree of correlation between the timeframes given in the two accounts.[2] It is also evident that some of these texts, such as the Gilgamesh Epic and SKL, also remembered the Noahic Flood as a real, historical event. The SKL also mentions similar characters to the Biblical account where, according to David Rohl, Meskiagkasher may be identified as the Biblical Cush, and Enmerker as the Biblical Nimrod.[3]

In the third century BC, Berosus, a priest of Baal-Marduk, offered a translation of the SKL as an extended history of Babylon consisting of tens of thousands (myriad) of years. History itself was believed to have extended for around two million years in four phases. Writing in the

---

[1] See for instance Rohl, D., *Legend: The Genesis of Civilisation*, Century, 1998, pp. 164-5
[2] Cooper, W., *The Authenticity of the Book of Genesis*, CSM, 2012, pp. 59-63. Cooper refers to a paper by Lopez, R.E, 'The Antidiluvian Patriarchs and the Sumerian King List', *CEN Tech. J.* 12 (3), 1998, pp. 347-357
[3] Rohl, *Legend.*, pp. 164-165.

Babyloniaca, a document prepared for Antiochus I, around 280 BC, he asserted that ten anti-diluvium kings together had reigned for 432,000 years. However, he also recognised the reality of the deluge and the Flood hero Xisuthrus.[4] Berosus followed Babylonian convention and divided time into equal units where one saros was commonly believed to represent a period of 3,600 years, a neros 600 years, and a sossus 60 years. The ten anti-diluvium kings then reigned for 120 saros. But prior to that, in Berosus' first phase from creation to the ancient kings, there was a period of 466 saros and 4 neros; in total a period of 1,680,000 years. From the Flood to Alexander the Great there had been 10 saros or 36,000 years. He thought that from 323 BC to the end of time there would be a further 3 saros and 2 neros or 12,000 years. In total the history of the present cosmos would be 600 saros or 2,160,000 years.[5]

It is known that the Babylonians used a sexagesimal numbering system; that is one based upon multiples of the number 60. Knowledge of this may have been lost due to cultural misunderstanding through translation into Greek, although interestingly it correlates closely with the Hindu system, as will be discussed below. William Cooper also believes that there may be a smaller unit of saros of 18.5 from sources quoting the tenth century Greek scholar Suidas.[6] The 10 anti-diluvium kings though reigned for 432,000 years as follows in the Table 1 (below), according to Eusebius' understanding of Berosus.

Names in Berosus' SKL	Reign in saros	Estimated in years
Alorus	reigned for 10 saros	36,000 years
Alaparus	reigned for 3 saros	10,800 years
Almelon	reigned for 13 saros	46,800 years
Ammenon	reigned for 12 saros	43,200 years
Amegalarus	reigned for 18 saros	64,800 years
Daonus	reigned for 10 saros	36,000 years
Edovanchus	reigned for 18 saros	64,800 years
Amempsinus	reigned for 10 saros	36,000 years
Otiartes	reigned for 8 saros	28,800 years
Xisuthrus	reigned for 18 saros	64,800 years

---

[4] McCalla, A., *The Creationist Debate*, London: Continuum, 2006, p. 29
[5] These dates are sourced via Jaki, S.L. *Science and Creation*, Scottish Academic Press, Edinburgh, 1974, pp. 96-98
[6] Jones, F.A, *The Dates of Genesis*, Kingsgate Press, 1912 [first published 1909], p. 114. Quoted in Cooper, *The Authenticity of the Book of Genesis* p. 63.

This list covers a total period of 120 saros, or 432,000 years.[7] So the timeframe of the pre-Flood Patriarchs was extended way beyond those given in the Genesis account, and longer than the SKL. Eusebius of Caesarea offered a number of reasons why he thought the Babylonian timeframe might have been extended in this way. He suggests, by way of example, that the ancestors to the Egyptians used a lunar cycle where a 30-day period constituted one month, and was referred to as a 'year'. Elsewhere a three-month period was referred to as 'hours'. Eusebius then is suggesting that understanding how the Chaldeans used the *sars* may need revising.[8]

So, the numbers used in Barosus' system should not necessarily be read literally and it is likely that they symbolically represent a shorter period of time, although it is not important to try and resolve this further here. Another possible reason for extended timeframes may have been for reasons of national pride. Manethos, an Egyptian priest of Heliopolis, provided a record of the Egyptian kings grouped into dynasties with a total dynastic period of 36,525 years. Historian Arthur McCalla has commented that both Manethos and Berosus were seeking to show that Babylonian and Egyptian civilisations were superior to that of the conquering Greeks because of their lengthy history.[9] In other words, an extended history was important for reasons of national prestige thus revealing the human tendency to exaggerate over matters of national pride. However, it is possible that a measure of this historical extension was due to a failure to understand the way ancient civilisations calculated time when translated across cultures.

With the rise of Hinduism, hundreds of thousands of years were extended into millions or even billions of years with a rather depressing cyclical view of history. People were then bound to follow the same paths time and time again, and belief in a continual cycle of reincarnation and death meant that there was little hope of escaping from the hand of fate (a cyclical eternal history later appeared in the work of Aristotle). Stanley Jaki observes that such pessimism, involving long cyclical ages, was possibly due to traumatic experiences that the Asiatic people had suffered in their history. The *Vishnu Purana* sets out the division of time as follows in Table

---

[7] This list is taken from Eusebius' Chaldean Chronicle, Ch.3, Transliteration by Robert Bedrosian (2008), sourced from the 1818 Venice translation by Father Mkrtich' Awgerean (Jean-Baptiste Aucher). (See http://rbedrosian.com/euseb2.htm).
[8] Eusebius' Chaldean Chronicle, Ch.3
[9] McCalla, *The Creationist Debate,* 2006, p. 29

2 (below), and Jaki notes that there is also influence from animistic, pantheistic and cabbalistic sources in this schema.[10]

1 divine day		1 year
1 divine year		360 years
12,000 divine years	4 Yugas or 1 Maha Yuga	4,320,000 years
1000 Maha Yugas	1 kalpa (1 day or night Brahma)	4,320,000,000 years
1 day and night of Brahma	2 kalpas	8,640,000,000 years
1 year of Brahma		3,110,400,000,000 years
100 years of Brahma	One life of Brahma	311 trillion and 40 billion years

Frederick Jones has shown that there is a correlation between the Babylonian timeframe of Berosus and that of the Hindus with history divisible by units of 432,000 years in the ratio of 4, 3, 2 and 1. So that there were four ages of 1,728,000, 1,296,000, 864,000 and 432,000 years (giving a total of 4,320,000 years), the present age being an imperfect half-age called the Kuri Yuga.[11] The detail of this is not that important to us here, except to note that the sexagesimal counting system is evident in the Brahma Hindu cosmology, as with Berosus' Babylonian system. Jones, however, seeks to get to grips with the Hindu numbering system and suggests that a period of 864,000 years (two Kuri Yugas) may relate either to the number of tenths of seconds (*matires*) in a day, or to the number of hours in a century.[12] It is noteworthy that, given 360 days in a year, the number of

- tenths of seconds in a day (10 x 60 x 60 x 24) = 864,000
- hours in a century (24 x 360 x 100) = 864,000
- tenths of seconds in a year (864,000 x 360) = 311,040,000

The Hindu numbers then are highly specific sequences that relate to the number of seconds in a day and year based upon the number 60. While there may be disagreement between commentators over exactly how the Hindu timeframe relates to the Babylonian one, there does seem to be a

---

[10] Jaki, *Science and Creation*, 1974, pp. 3-5
[11] Jones, *The Dates of Genesis*, p. 114.
[12] Sourced via Maunder, E.W., 'Book Review: The Dates of Genesis,' *The Observatory*, Vol. 32, 1909, pp. 390-393

significant correlation, suggesting shared influence and a common source. The sexagesimal system is present in both. And such correlation between Hinduism and Baal worship should not surprise us when we may note that both worship the bull as a sacred animal. Some caution does need to be made in claiming that the ancient Hindu's intended billions of years in their cosmology, or whether it was meant to indicate periods involving much shorter timeframes.

In the Hindu Upanishads the world itself is said to be evolving, or unfolding as a self-creation (or pro-creation) where the whole world is viewed as the god Brahma. Berosus also considered that life arose through a process of generation or evolution from simpler organisms. A number of odd creatures were seen emerging and evolving out of the oceanic chaos, such as fish and dogs with human heads, or various reptiles and snakes crawling out of the sea. Some of these were able to reproduce asexually, while others crossbred to produce new forms.[13] Such Babylonian beliefs in the generation of life and the gods may later have influenced Greek thought. This may be seen for instance in Hesiod's *Theogony*, with the Babylonian *Theogony of Dunnum* possibly being influential.[14] Hindu and Babylonian polytheism further gave the universe, and the world, a soul as a self-creating or unfolding entity. With Buddhist influence, the Chinese of the Han Dynasty (206 BC to 220 AD) also saw their history as being cyclical and extending to millions of years with influence arising from conjunctions of planets, and eclipses of the moon and sun. Using a metonic cycle, one consisting of 19 years or 235 lunations, the Chinese Han astronomers developed a complex cyclical history, as stated in the *Chou Pei Suan Ching*, with a grand period of 31,920 years, and a world cycle of 23,639,040 years.[15]

Jaki comments that such was the similarity with Hindu and Buddhist cosmologies that the Chinese neo-Confucians were happy to assimilate the Hindu cyclical history into their own. All were strongly pantheistic or animistic; that is the belief that the celestial bodies possessed souls.[16]

Although this evidence from the *Babylonica* and Hinduism post-dates Moses, we may extend this back in time. The ancient nations that

---

[13] See The Chaldean Chronicle of Eusebius, Ch.5

[14] Lambert, W.G, and Walcot, P. 'A New Babylonian Theogony and Hesiod', *Kadmos* 4 (1), 1965, 64–72

[15] Jaki, *Science and Creation*, 1974., pp. 33-34 (1 *chang* equals 19 years. 4 *chang* equals 1 *Pu* of 76 years. 27 *chang* equals 1 *hui* of 513 years (or 47 lunar eclipse periods). 3 *hui* equals 1 *thung* of 1,539 years. 20 *pu* equals 1 *chi* of 1,520 years. 3 *chi* equals 1 *shou* of 4,560 years. 7 *shou* equals the Grand Period of 31,920 years).

[16] Jaki *Science and Creation*, 1974., p.34

surrounded the emerging nation of Israel, as it came through the Sinai Desert, may in fact have believed in histories extending back millions of years. However, the difficulty for us here is getting to grips with exactly how these nations developed their understanding of time, and exactly when they adopted millions of years of history. The extension of timeframes to millions of years may have been due to misunderstandings in translation, or for reasons of national pride, or both perhaps. There does seem to have been an original common source between the Sumerian King List and the Genesis account, and even with later corruptions between the Babylonian and Hindu belief systems.

Genesis as a Polemic against Paganism

The account of Creation that came through Moses was short in comparison with the Babylonian and Hindu timeframe and it was careful not to idolise the world as the polytheistic accounts were doing. Ernest Lucas has pointed out that the Genesis account may be seen as a 'polemic' against the pantheistic or polytheistic cultures of the ancient Middle East. He gives as an example the creation of the sun and moon as lights placed in the sky by God in contradiction of the pagan tendency to worship them as divine objects.[17] Other examples given by Lucas are the divine creation of the sea monster (known as the god *Tiamat* by the Babylonians) that speaks of God's control over the forces of chaos, and the creation of man as having a status above that of slaves. Genesis should then be seen as an account that seeks to establish monotheism and a divine creation, the purpose of this to educate the people of Israel and train them to worship the true creator God in the face of polytheism and idolatry. We may further observe that the polytheistic accounts of creation seek to deal with ideas present in Genesis, although in a far less literal and rational manner than the Mosaic account. Examples are the presence of chaos, of bringing form out of the chaos, and the existence of good and evil, night and day and light and darkness.

While Lucas offers some important insights here we may take his argument further and ask a couple of pertinent questions. The first question relates to the relatively short history of the world according to Genesis, together with God's direct work in creation, and his progressive plan for his people. In the same way that the sun and moon are described as created objects the Genesis account stresses a recent creation *ex-nihilo*, this against the developing pagan beliefs in long ages, creation out of a pre-existing

---

[17] Lucas, E, 'Science and the Bible, Are they Compatible?,' *Christians in Science - St Edmunds College Public Lecture*, 13th May 2004 (also *in Science and Christian Belief* (2005) Vol. 17 pp. 137-154). Lucas quotes Hasel, G., 'The Polemic Nature of the Genesis Cosmology', *Evangelical Quarterly*, 46, 1974, 81-102.

chaos, and a cyclical view of history. The polytheistic cultures often held to belief in an impersonal *god-like* power of generation in nature, and even to the generation of the gods (*Theogeny*). The God of the Bible instead is presented as the Creator of all things, one who is interested in his people and one who acts decisively. That is a dynamic, intelligent and personal agent, instead of belief in a slow acting mystical and impersonal force or power at work in nature. The account of a recent and direct creation in Genesis then may be read as statements of fact of what God has done; that is, if the account is to be read as a polemic against polytheism and pantheism.

Secondly, Lucas suggests that the Genesis account should be seen as 'accommodated' to the simple understanding of the people and therefore it does not reveal science. He refers to Calvin's commentary on Genesis to make his point.[18] Of course the Mosaic account is not interested in detailed science. For instance it does not tell us the size of the earth, or the distance to the moon or sun (although it does seem to speak about time). Neither does it tell us anything about the structure of the atom, but it would be wrong to see Genesis as a simple document. In fact, in the Genesis passage, detailed science is considered to be of far less importance than the spiritual development of God's people. Genesis may however be seen as describing historical events concerning the formation of the world in a rational manner.

Furthermore, the symbolism present is also highly structured and complex, revealing theological truths that are anything but simple. It deals with the important question of suffering for instance. And Beale suggests the account of Creation and the Garden of Eden should be read in the context of the foundation of a universal temple (in which mankind may worship God) and that this provides background information to the giving of the Mosaic Law that include temple practices.[19] The Law of Moses is also highly detailed and provides the basis for an ordered society. But just because the account of creation in Genesis is not interested in detailed science, as we know it today, it does not mean that it is a simple document written for simple people, and that it is not interested in real history.

In view of the above discussion we may consider a further point. If the Genesis narrative was actually simplified to accommodate its understanding for simple people, i.e. by stressing a recent miraculous event when God had *really* created over long periods of time through an evolutionary process, then we may wonder why the text would not simply state that as so? This is because beliefs involving deep time and forms of

---

[18] Calvin, J. *A Commentary on Genesis*, King, J. (trans.), London: Banner of Truth, 1967. (Lucas refers to commentary on Genesis 1:5-16).

[19] Beale, G.K. *The Temple and the Church's Mission*, IVP, 2004

evolutionary thought were developing in several of the surrounding idolatrous cultures. So, the congregation of Israel could just have easily understood an evolutionary account and deep time, instead of one involving a recent direct creation by God, *if* simplicity were really the criterion and *if* evolution was really the way creation had arisen. If God had really created through an evolutionary progression and long ages, then why not just say so? It would have at least been understandable to the people. The fact that the account does not teach that is because God did not create in such a way, and it would be wrong for us to read that into the text by arguing for deep time and theistic evolution. If the Genesis account is a polemic against pantheism and paganism, because it sees the sun and moon as real created objects, then it is also a polemic against deep time and molecule-to-man evolutionary powers of generation.

Evolution and Long Ages in Ancient Greek Thinking

As noted the ancient Hindus believed that the world itself was evolving due to an impersonal force at work in nature. The concept of an evolutionary progression can also be traced back through ancient Greek thinking. And there does appear to be a measure of commonality between the beliefs of the ancient civilisations of the Ancient Middle East and Far East. Michael Denton has provided some background to the development of evolutionary ideas in Greek antiquity starting with the pre-Socrates materialists.[20] He follows the path back via David Hume to the Greek materialists such as Epicurus (341-270 BC), Democritus (460-370 BC), and to Anaximander of Miletus (610-546 BC). Anaximander, who was perhaps the pupil of Thales, believed that all life had a material origin being derived from sea slime. This is echoed in more recent theories of life evolving out of primordial soup. Empedocles, by 450 BC, believed that nature was always throwing up oddities, or 'hopeful monsters', which would either live or die according to their suitability (and as such it reflected Darwin's later idea of natural selection). This led to the Atomists of the fifth century BC to claim that the gods were no longer necessary. However, atheism was not universally accepted in pagan Greece, even being illegal at times. The idea that impersonal forces may be at work in nature appears to suit the atheists and pantheists alike with a subtle dividing line between them, as the pagan concept of nature seems to display ambiguity.

This ambiguity is also seen in David Hume's *Dialogues Concerning Natural Religion,* which had an important influence on both the evolutionary theories of Erasmus Darwin and later his grandson Charles

---

[20] Denton, M, *Evolution: A Theory in Crisis*, Bethesda, Maryland: Adler and Adler, 1982.

Darwin. In Hume's *Dialogues*, as well as the Epicurean perspective, a pantheistic perspective is offered through his character Philo (in Part VII). Subsequently, in later sections, he offers the more atheistic Epicurean perspective, also through the character Philo. In the *Dialogues* Philo is generally seen as speaking for Hume against the classical design argument, but the script is complicated by Philo's equivocation (the other characters are Cleanthes, who is the design proponent, and Demea who argues that reason and divinely given feelings can establish the truth of religion). Philo, as arguably the mouthpiece of Hume, at first stresses his scepticism of all cosmologies, but Philo does in fact also suggest that he would prefer the pantheistic position of Part VII of the *Dialogues*, if he were forced to choose, as opposed to the Epicurean one. Hume, through Philo, claims that such a system is sourced from the 'ancient mythologists' such as Hesiod and Plato, and also from the Brahmins who believed the world arose from an infinite spider.

> Hesiod, and all the ancient mythologists…universally explained the origin of nature from an animal birth, and copulation. Plato too, so far as he is intelligible, seems to have adopted some such notion in his Timaeus.[21]

Hesiod was a Greek poet who lived around 700 BC and is often associated with Homer. One of the works attributed to Hesiod is the poem *Theogony*, which is concerned with the origins of the world and of the gods - the title means the 'generation of the gods.' According to Hesiod the first god to arise was *Chaos* who was said to have arisen spontaneously, followed by *Gaia*, *Eros* and *Nyx*. A whole pantheon of gods, elements and living animals were then generated and modified through descent.

In the *Dialogues*, according to Hume's character Philo, a tree sheds its seeds into the surrounding fields in order to produces other trees, the 'great vegetable' that is the world or planetary system also produces seeds, which are scattered into the chaos, and these then vegetate into new worlds. Philo suggests that a comet might be the seed of a world, or the world might be an animal, and a comet the egg of an animal. Philo argues that generation and reason are words that 'mark only certain powers and energies in nature, whose effects are known, but whose essence is incomprehensible.' And further asserts that there are four principles known on the earth, which are 'reason, instinct, generation [and] vegetation,' but

---

[21] Hume, D. (1947) 'Dialogues Concerning Natural Religion,' in Kemp Smith, N., (ed.) *Dialogues Concerning Natural Religion*, 2nd ed. Indianapolis: Bobbs-Merrill Educational Publishing, p.180, the text used here is from *The philosophical works of David Hume*, 1854.

that there may be other unknown principles in the cosmos.[22] As noted Philo argues that order need not spring from thought.[23] Philo is then comparing his argument on this pagan principle of generation against Cleanthes' more classical design argument. Hume writes, through Philo, that 'The world, say I, resembles an animal; therefore it is an animal, therefore it arose from generation.' On the other hand Cleanthes' argument is that the world 'resembles a machine; therefore it is a machine, therefore it arose from design.'[24]

So the idea of a power or source of generation, as an early concept of evolution, can be traced back in Greek thought to Plato's *Timaeus* and Hesiod's *Theogony*, and perhaps to the Hindu Brahmins as well. Plato in fact presented the idea of a designer in the *Timaeus* as the Demiurge, a skilled craftsman. The name *Demiurge* (*demos* = common people, *ergon* = work) was initially used for a human craftsman or skilled worker such as a stonemason or architect, but was later used to refer to the lesser god who fashioned the universe out of a pre-existing chaos. The Demiurge was said to have created through three integral aspects in the universe. These are the *arche* as a source of all things, the *logos* representing a hidden order, together with the harmony of mathematical *ratios*. Plato thought the world was patterned on ideas that had existed in the mind of the Demiurge.

Hume's writing though leaves us with a problem that needs an answer. Why does Hume appeal to a work in which we find the concept of a cosmic designer, the Demiurge, when he seems to use it to undermine that very same concept by attacking Cleanthes' design argument? The solution comes by understanding Plato's worldview and metaphysics, and the desire in Hume's writing to detach design from an analogy to human intelligence. This was taken one stage further in Charles Darwin's *The Descent of Man*, where mankind no longer has a place as God's special creation, but is merely a bi-product of a random evolutionary process. In pagan thought there is a tendency to replace the personal and relational Judeo-Christian God with an impersonal force, or lesser gods that are not interested in human affairs. Such impersonal forces do not make demands upon people, but leave people unaccountable with the apparent freedom to shape their own destiny, but ultimately it is a hopeless and unloving existence.

For the Greek thinkers such as Plato, natural objects within the real world were mere shadows of a greater spiritual reality. The body was considered corrupt with the soul released upon death into the 'good.' Although the Demiurge wanted the world to be as perfect and as good as possible, it was imperfect because the Demiurge could only make the

---

[22] Hume, *Dialogues*, p. 178
[23] Hume, *Dialogues*, p. 179
[24] Hume, *Dialogues*, p. 180

universe out of the chaos (and this is reflected in dualistic Gnostic heresies). Because the world was seen as imperfect, the task of human beings was to look forward to the liberation of the soul from the corruption of the world and body, and ascend through the heavens upon death. In the meantime the purpose of contemplation and philosophy was to understand the divine mind and ease suffering by turning a blind eye to it. This led to a strong dualism between spirit and matter, which is reflected in the modern secular belief that science and faith must be kept completely separate.

David Sedley comments that the pagan religions never saw the act of creation by a divine agent as being out of nothing as Christian theology taught, but always out of matter in the pre-existing chaos. This suggested an eternal, cyclical and evolving cosmos. Neither did the pagans have much interest in the literalness, or authority of religious texts, as is the case with Judeo-Christian reading of Moses.[25] Instead the allegorical meanings were considered superior. Sedley relates this to the difficulty faced by scholars in interpreting Plato, and argues that his works ought to be read in an allegorical, or metaphorical manner as well.[26] And this helps to shed light on Hume's comments above relating to Plato's *Timaeus* and Hesiod's *Theogony*. It helps us to understand what Hume was implying about the metaphorical nature of the designer as an impersonal force or power of generation, i.e. a pantheistic form of evolution.

It has been shown then that Hume traced the concept of generation as a source or power in nature back to pagan beliefs, from Hesiod and Plato and from the Hindu Brahmins. The concept of generation in ancient pagan thought provided both for the continuation of life forms through time, together with the possibility of change; i.e. modification with descent as an early form of evolutionary belief. As will be discussed in greater depth in subsequent chapters, this is a phrase that is similar to comments by Erasmus Darwin in *Zoonomia* who believed in a 'power or source of generation' linking it to the 'Great Architect,' i.e. the Demiurge, who 'makes the maker of the machine.'[27] Quite clearly Charles Darwin's view was much more subtle and closer to the more atheistic Epicurean perspective, but the ambiguity in the concept of nature remains.

---

[25] Sedley, D., *Creationism and its Critics in Antiquity*, LA: University of California Press, 2007, p. xvi, xvii

[26] Sedley, D., *Creationism and its Critics in Antiquity,* pp. 98-107.

[27] Darwin, E. *Zoonomia; or the laws of organic life*, Vol. 1, New York, 2nd American Ed, from 3rd London Ed., corrected by the Author, Boston Thomas and Andrews. 1803: pp. 400-401

Chapter summary

It would seem then that the writings of Moses were revealed to the Israelites when belief in deep time and evolutionary change were already developing in the polytheistic nations that surrounded them. The idea, as claimed by some theistic evolutionists, that God had to accommodate the language of Genesis because the Israelites could not accept the idea of long ages and evolutionary change is therefore untenable. Genesis may be seen as a direct revelation, perhaps as a definitive statement, to counter the pagan beliefs that the Israelites had picked up in Egypt or elsewhere. From this we can see that the account should be read both symbolically and literally, and as will be shown in the next chapter, this is how the Church Father's understood it. The Creation account places mankind as God's direct and recent work, given a special place to love God and to be in relationship with him, and also to have responsibility to care for the rest of creation. God not only fine-tuned laws of physics and chemistry, but he fashioned Adam from the dust of the ground, giving him three-dimensional form, shape, and life. Furthermore, human kind was not made merely as a machine or as slaves of the gods, but as beings made in the image of God for a transcendent relationship.

All of the seeds for acceptance of long ages and belief in an evolutionary progression were present in ancient times; being derived from pagan beliefs as far a field as India, Egypt, Babylon and Greece. Modern cosmology is in fact remarkably close to the timeframe of one day of Brahma. Is that merely a coincidence bearing in mind how much influence pagan beliefs have played in the development of natural science? It is not merely a coincidence, and this can be supported with evidence, as will be show through later chapters. In other words, natural science, involving deep time and evolution, is not free from its own polytheistic or pantheistic religious influence; this despite pronouncements from secularists that science should be free from such influence. The concepts of nature and evolution can be traced back to ancient times, where for instance Hesiod believed that everything was caused by a process of generation from the gods downwards, as a sort of modification with descent; and Plato seems to have developed a similar scheme if he is read metaphorically, as leading commentators suggest he should be read. While it is true that the Greek materialists abandoned the pagan gods, as has modern Darwinian science at face value, the pagan beliefs provided the central foundation for their thinking and formation, and there remains ambiguity in the idea of nature between pantheism and atheism. Nature is often ascribed god-like power because it has a creative force over itself and is self-existing. The boundary is often blurred, a recent example being Richard Dawkins *The Magic of*

16

Reality. This is something that ought to be of concern to theistic evolutionists as well.

Charles Darwin seemingly abandoned the pagan considerations of his grandfather, but belief in deep time and evolution was developed out of pagan beliefs even so. In other words, ancient pagan beliefs are the foundational inspiration for belief in deep time and the grand evolutionary process that is beloved by natural scientists. Through the writings of Erasmus Darwin and David Hume, modern evolutionary ideas are based on the pagan belief of Hesiod and Plato that arose from Babylonian and Hindu beliefs. In this, there is considered to be a power or source of generation at work in nature that can account for all living things, and mankind is merely a part of that process. The irony is that these polytheistic or pantheistic beliefs are themselves developed corruptions from an earlier more rational knowledge that was monotheistic and recognised a recent direct Creation and a damaging Flood.

3.

The Church Fathers, Creation and Science

In the context of this book it is relevant to discuss the beliefs of the Church Fathers in relation to their Christian faith, and the Greek culture in which they lived. Largely, they believed that the earth was historically less than 6000 years old, even though we may also note that they often read the Genesis account symbolically as well. In places they also adopted Aristotelian science, with its commitment to geocentrism and spontaneous generation, but some were critical of the spiritual-versus-material dualism of Platonism, which was seen as a form of Gnosticism. Augustine for instance tried to hold in balance the literal and symbolic readings of Scripture. On the other hand, a few Christian leaders, such as Origen, were more focussed upon the symbolic reading of nature.[1]

However, we may note that today some theistic evolutionists and progressive creationists use the Church Fathers to try and support belief in deep time and evolution. Denis Alexander for instance quotes Augustine to claim that the creationist position is an embarrassment to the gospel.[2] John Lennox suggests that Irenaeus and Justin Martyr may have been sympathetic to an older earth, with the six-days seen as 'long epochs' of time.[3] He seems to be following Hugh Ross here despite recent work that shows these theologians held to a creation-millennial scheme because of their interest in the timing of the return of Christ.[4] Furthermore, Lennox

---

[1] This chapter is based in part upon a previous published paper; Sibley, A, 'Creationism and millennialism amongst the Church Fathers', *Journal of Creation*, 26(3), 2012. I have made use of Philip Schaff's *Anti-Nicene Fathers*, (10 vols. first published 1885), and *Nicene and Post Nicene Fathers,* (28 Vols. Series I and II, 1886-1890) available online at www.ccel.org

[2] Alexander, D, *Evolution or Creation: Do We Need to Choose?* Monarch, 2009, pp.351-353. Mention of individuals here is offered in a spirit of respectful dialogue.

[3] Lennox, J., *Seven Days that Divide the World*, Grand Rapids: Zondervan, 2012 pp. 30-33 & 40-42. I have greatly appreciated Lennox's ministry over the years.

[4] Ross, H., *Creation and Time: A Biblical and Scientific Perspective on the Creation-Date Controversy.* Colorado Springs, CO: Navpress, 1994, pp. 17-19. James Mook shows that Irenaeus and Justin Martyr were proponents of a millennial scheme; Mook, J.R., 'The Church Fathers on Genesis, The Flood, and the Age of the Earth', in Mortenson, T., Ury, T.H., (eds) *Coming to Grips with Genesis*, Green Forest AR, 2008. Also discussed by Robert Bradshaw, *Creationism and the Early Church*, 25 January 1999 http://www.robibradshaw.com/contents.htm (accessed December 2012). See also, James Millam, 'Coming to Grips with the Early Church Fathers' Perspective on

believes young earth creationists fall into a similar error to those who held to geocentrism. However, this really overlooks the beliefs of the early theologians, as well as the biblical origin of one versus the pagan origin of the other. Roger Forster and Paul Marston also address the beliefs of the Church Fathers, but downplay the significance of their literal reading of Genesis. Where they do address literal readings it is in rather mocking tones. They also seek to break the link between belief in a young earth and Christian tradition, although this really fails in light of evidence.[5]

From a close reading of the Church Fathers it becomes apparent that a widely held belief was that the Creation occurred less than 6,000 years ago; perhaps not exclusively so, but it would seem to have been a majority opinion. This belief tied in with the question of the timing of the Second Advent, which was of prime importance to them. The millennial scheme then provided a major framework to try and calculate the timing of Christ's return. This symbolically linked the six days with 6,000 years of earth history, with the final millennium rest seen as foreshadowed by the seventh day of rest in the Creation account.

Literal and Allegorical Readings

It is true that a number of the Church Fathers, such as Origen, emphasised the allegorical over the literal reading of the Genesis account, and Forster and Marston highlight the ambiguity in Origen's thoughts in this matter.[6] But there was a desire amongst others to hold in balance both the literal and allegorical readings. Basil, the Cappadocian saint who lived from 330 to 379 AD, acknowledged the 'laws of allegory,' but also emphasised the literal sense in his interpretation of Genesis. In his *Hexaemeron* (meaning *Six Days*) he wrote.

> I know the laws of allegory, though less by myself than from the works of others. There are those truly, who do not admit the common sense of the Scriptures, for whom water is not water, but some other nature, who see in a plant, in a fish, what their fancy wishes, who change the nature of reptiles and of wild beasts to suit their allegories, like the interpreters of dreams who explain visions

---

Genesis,' Parts 1-5 http://www.reasons.org/articles/coming-to-grips-with-the-early-church-fathers-perspective-on-genesis-part-1-of-5, 8 September 2011. (Accessed December 2012).

[5] Forster, R & Marston, P, *Reason, Science and Faith*, Monarch, 1999, pp. 200-201, & 237, 241.

[6] Forster and Marston, *Reason, Science and Faith*, pp. 203-204, referring to Origen *Homily I* and *First Principles*, IV: 3.

in sleep to [m]ake them serve their own ends. For me grass is grass; plant, fish, wild beast, domestic animal, I take all in the literal sense. "For I am not ashamed of the gospel."[7]

For Basil symbolism was recognised in the biblical narrative, but he also upheld the importance of the literal account, including in terms of the length of a day. He saw that in Genesis a day was a period of 24 hours.

And there was evening and there was morning: one day." And the evening and the morning were one day. Why does Scripture say "one day the first day"? Before speaking to us of the second, the third, and the fourth days, would it not have been more natural to call that one the first which began the series? If it therefore says "one day", it is from a wish to determine the measure of day and night, and to combine the time that they contain. Now twenty-four hours fill up the space of one day—we mean of a day and of a night.[8]

It is questionable though whether Basil held Greek science in high esteem, even though engaging reasonably well with it, for instance when discussing the size and shape of the Earth he suggested that different natural philosophers contradict one another. Therefore Greek philosophers should not mock Christian beliefs until they are settled themselves in what is true.[9] But where he did engage with the science of his time it was in Aristotelian terms that we would find difficult to accept, i.e. seemingly accepting geocentrism, and spontaneous generation where the earth may for instance produce eels directly from the ground.[10]

Augustine, who lived from 354 to 430 AD, similarly held to a form of spontaneous generation with his idea of the *rationes seminales*, but it was action in response to the divine word.[11] Potential divine words as

---

[7] Basil, *Hexaemeron*, Homily, IX:1

[8] Basil, *Hexaemeron*, Homily, II:8

[9] Basil, *Hexaemeron*, Homily, III:3, and II:8. Basil wonders why the Christian belief in several heavens should be seen as foolish when the Greek are willing to propose 'infinite heavens and worlds.'

[10] See for instance Basil, *Hexaemeron Homily* II:8, "in reality a day is the time that the heavens starting from one point take to return there. Thus, every time that, in the revolution of the sun, evening and morning occupy the world", and in *Homily* IX:2, "We see mud alone produce eels; they do not proceed from an egg, nor in any other manner; it is the earth alone which gives them birth. Let the earth produce a living creature."

[11] See Augustine's, *De Genesi Ad Litteram* , Book 3.14.23 & Book 4.33.51, translated and annotated by Taylor, J.H., Newman Press, New York,1982.

'seeds' were planted in the ground at the beginning, but actualised through time as new plants, or small animals continually arise out of the ground. Interestingly, this Augustinian belief may have influenced continued belief in spontaneous generation, and belief that fossils formed in the ground through a plastic force of nature, in the early modern period.

On the question of accepting a 'hard roof' dome-like firmament over the earth Basil compared analogically the strength of the firmament to that of the strength of the air when exposed to thunder. From Isaiah he maintained that the heavens 'are a subtle substance, without solidity or density.'[12] He further comments that

> It is not in reality a firm and solid substance which has weight and resistance...But, as the substance of superincumbent bodies is light, without consistency, and cannot be grasped by any one of our senses, it is in comparison with these pure and imperceptible substances that the firmament has received its name.[13]

There is a sense then that for Basil the literal reading concerning the starry heavens retained a spiritual aspect, a way of viewing creation that is lost with the dominance of materialism in science, and is sometimes overlooked by those committed to creation science as well, especially in relation to understanding the pre-Fall world. For Basil, the inspired writings of Moses were given to provide spiritual truths, and these were considered more important than exact measurements and distances that are not of interest to the Genesis account. However, he accepted the times and dates given in an historical manner. The works of Moses were also seen as of greater worth than the writings of the Greeks and therefore took precedence over them.

Augustine also interpreted Scripture in a way that included the literal, but also allegorical or symbolic readings that pointed to spiritual truths. He recognised that there was an overlap between both types of readings with the symbolic arising from the lives of real people and events. Concerning Noah's Flood Augustine held together the literal and allegorical and wrote in a chapter in *The City of God* that;

---

[12] Forster and Marston, *Reason, Science and Faith*, pp.200-201. They merely quote, without a page reference, from Stanley Jaki, *Genesis 1 through the Ages*, 2nd ed. Scottish Academic, 1998. The reference to air and thunder in Basil I have sourced from *Hexaemeron* Homily III:4, the reference to heaven and smoke is from Homily I:8, but Basil commenting further that 'He created a subtle substance, without solidity or density, from which to form the heavens.'

[13] Basil, *Hexaemeron* Homily III:7

...no one ought to suppose either that these things were written for no purpose, or that we should study only the historical truth, apart from any allegorical meanings; or, on the contrary, that they are only allegories, and that there were no such facts at all....[14]

Augustine's approach held in balance the literal truth and the symbolic meaning. It followed from the disagreement between the Alexandrian School, particularly of Origen, that emphasised the allegorical above the literal, and the Antiochene School that focussed upon reading the prophetic writings primarily in the historical context.[15] As such there was reluctance to see Christ in the Old Testament texts within the excessively literalistic framework. Augustine though saw symbolism relating to Christ and the cross in the account of Noah and the Ark. The Augustinian approach then sought to blend together the theological and symbolic aspects, where we can see Christ in the Old Testament as well as the New. And where we see our faith having a real impact upon the material world, including through the study of creation. It is this combined literal-symbolic reading that allowed the early Church to infer 6,000 years of Earth history from the six days of Creation.

The Age of the Earth and Six Days of Creation

Influence for the symbolic linkage between the six literal days and the 6,000 years until the millennium reign of rest may have been passed down directly from the apostles, especially John. Irenaeus, who lived during the second century AD, claims to have received the teaching from Papias and Polycarp who he suggests received it directly from John.[16]

---

[14] Augustine *The City of God*, Book XV, Ch 27; entitled 'The Ark and the Deluge, and that We Cannot Agree with Those Who Receive the Bare History, But Reject the Allegorical Interpretation, Nor with Those Who Maintain the Figurative and Not the Historical Meaning.'

[15] McGrath, A.E, (2002) *Christian Theology: An Introduction*, 3rd edition, Blackwell, pp. 171-178.

[16] Irenaeus, *Against Heresies* 5:XXXIII:4, "And these things are borne witness to in writing by Papias, the hearer of John, and a companion of Polycarp, in his fourth book; for there were five books compiled (συντεταγμένα) by him." And in *Against Heresies* 5.XXX:4, "then the Lord will come from heaven in the clouds, in the glory of the Father ...bringing in for the righteous the times of the kingdom, that is, the rest, the hallowed seventh day". Also see Papias, *Fragments* ch. IX (sourced via Anastasius Sinaitia), which reads, "Taking occasion from Papias of Hierapolis, the illustrious, a disciple of the apostle who leaned on the bosom of Christ, and Clemens, and Pantænus the priest of [the Church] of the Alexandrians, and the

Influence also seems to have derived from Peter's writing, wherein the apostle was responding to questions about the return of Jesus Christ. Peter wrote that for God 'a day is like a thousand years' (2 Peter 3:8) and that the Christians should be patient regarding the question of Christ's return.

Reference to the notion that a 'day is as a thousand years' is also found in the writing of Justin Martyr (100 to 165 AD), although Justin used it in a slightly different way to others such as Irenaeus and instead linked it to passages from Isaiah. In his *Dialogue with Trypho a Jew* he asserted that there will be a millennial reign from a renewed Jerusalem and that the lives of the saints will be sustained through the thousand years. He quoted from the prophets Ezekiel and Isaiah, and John in Revelation. He also obscurely linked this to the age of Adam at his death.

> For as Adam was told that in the day he ate of the tree he would die, we know that he did not complete a thousand years. We have perceived, moreover, that the expression, 'The day of the Lord is as a thousand years' is connected with this subject.[17]

The implication of this is that the curse of death upon Adam would not be immediate, but within a period of one thousand years; i.e. one 'day' because Adam was told he would die in the 'day' he ate of the fruit. Adam lived 930 years. Some may of course suggest this supports the day-age scenario, as Ross, and Forster and Marston do,[18] but amongst the early Christian writers the millennial scheme of 6,000 years of earth history was clearly linked to the six days of Creation. As noted, the Creation and millennial timeframe is set out in the writing of Irenaeus. He held that the Earth was literally less than 6000 years old, and that this relates symbolically to questions about the return of Christ and the fulfilment of history. After 6,000 years Christ would return and reign for one thousand years, representing the Sabbath day of rest. Irenaeus wrote that

> For in as many days as this world was made, in so many thousand years shall it be concluded. And for this reason the Scripture says: 'Thus the heaven and the earth were finished, and all their adornment. And God brought to a conclusion upon the sixth day the works that He had made; and God rested upon the seventh day

---

wise Ammonius, the ancient and first expositors, who agreed with each other, who understood the work of the six days as referring to Christ and the whole Church."
[17] Justin Martyr, *Dialogue with Trypho*, 80-81.
[18] Forster and Marston, *Reason, Science and Faith*, 1999, p.201, Lennox, J., *Seven Days that Divide the World*, 2012, pp. 30-33 & 40-42, Ross, H., *Creation and Time*, 1994, pp. 17-19.

from all His works.' This is an account of the things formerly created, as also it is a prophecy of what is to come. For the day of the Lord is as a thousand years; and in six days created things were completed: it is evident, therefore, that they will come to an end at the sixth thousand year.[19]

It is apparent then that this is different to the day-age interpretation that many old earth creationists hold from the above passage in Peter. Similarly, Theophilus, the second century Bishop of Antioch (169 to 177 AD) agreed that the earth was less than 6,000 years old. He wrote that

All the years from the creation of the world amount to a total of 5698 years, and the odd months and days... For if even a chronological error has been committed by us, of, e.g., 50 or 100, or even 200 years, yet not of thousands and tens of thousands, as Plato and Apollonius and other mendacious authors have hitherto written. And perhaps our knowledge of the whole number of the years is not quite accurate, because the odd months and days are not set down in the sacred books.[20]

Similarly, we find in the fragments of Hippolytus of Rome a belief that the creation week occurred in the recent past at approximately 5,500 BC.

For as the times are noted from the foundation of the world, and reckoned from Adam, they set clearly before us the matter with which our inquiry deals. For the first appearance of our Lord in the flesh took place in Bethlehem, under Augustus, in the year 5500; and He suffered in the thirty-third year. And 6,000 years must needs be accomplished, in order that the Sabbath may come, the rest, the holy day "on which God rested from all His works." For the Sabbath is the type and emblem of the future kingdom of the saints, when they "shall reign with Christ," when He comes from heaven, as John says in his Apocalypse: for "a day with the Lord is as a thousand years." Since, then, in six days God made all things, it follows that 6,000 years must be fulfilled.[21]

Part of Hippolytus' justification was concerned with the size and adornment of the ark of Moses placed in the tabernacle, which was covered

---

[19] Irenaeus, *Against Heresies,* 5:38:3.

[20] Theophilus *to Autocylus*, 3: 28-29

[21] Hippolytus, Fragments - *Hexaemeron*, On Daniel II:4

in gold inside and out and measured in height, width and breadth, a total 5.5 cubits. Hippolytus believed that this distance signified the time of Christ's coming, and the ark itself 'constituted types and emblems of spiritual mysteries' that signified Christ and his coming. From this then Hippolytus calculated that 500 years remained until Christ returned and thus bring in the final Sabbath rest. 'From the birth of Christ, then, we must reckon the 500 years that remain to make up the 6,000, and thus the end shall be.'[22]

Many of the Church Fathers of the first three centuries relied upon the Septuagint (LXX) to determine the age of the earth and the second coming of Christ. This text was written several centuries before Christ by Alexandrian scribes, and gave an age of the Earth around 5,500 BC. Christians then believed that Christ would return sometime around 500 AD. However, the Masoretic Text (MT), with its shorter time frame of 4,000 years BC, slowly became accepted as the standard version of the Old Testament in later centuries. The MT then suggested a date of the Second Advent near to 2,000 AD. Jewish Scribes and Pharisees derived this standardised Hebrew Old Testament text from older versions around the time of the Council of Jamnia.[23] It is possible then that both the MT and the LXX differ somewhat from the test that Jesus and the Apostles may have had access to. The Samaritan Text differs also from both the MT and LXX.

In Christian circles the transition from the LXX to the MT began as early as the third century. Origen's desire was to get to grips with the differences between the LXX and the MT, and to aid his study he produced a six-fold interlinear version known as the *Hexapla*. However, the MT was brought more fully into the centre of Latin Christendom following Jerome's papal commission from Pope Damasus I in 382AD.[24] Jerome's Vulgate version became accepted in the Roman world for many centuries. The older Latin versions, based upon the LXX, fell into disuse. And with the Reformation the King James Bible also relied upon the MT for the basis of

---

[22] Hippolytus, Fragments – *Hexaemeron*, On Daniel II:5-6. Julius Africanus in the third century also held that the earth was around 5500 years old. He wrote that; 'The period, then, to the advent of the Lord from Adam and the creation is 5531 years' (*Chronography* 18:4).

[23] Sometime between 70AD and 135AD. This was for the reason of bolstering Jewish adherence, and to distance Judaism from Christianity. Jewish Christians had fled Jerusalem in AD 68 for Pella (following Jesus' prophecy relating to the sign of the abomination that causes desolation; see: Matt 24:16) and which increased distrust between Christians and Jews.

[24] Augustine maintained that the LXX was the more reliable version, although he found the many different Latin versions translated from the LXX to be frustrating, *On Christian Doctrine*, 2:16

its translation, and the MT informs most modern versions as well. The Greek Orthodox Church does however continue to use the LXX.

In terms of the age of the earth, Whitcomb and Morris believed that at least some of the dates in the LXX should be seen as false, even though they were open to a slightly longer timeframe for the age of the earth. For instance several thousand years might have passed between the Flood and Abraham.[25]

Augustine continued to favour the LXX, and held that the earth was less than 6,000 years old (although he apparently abandoned belief in a literal millennial reign of Christ on earth in later life). He wrote

> Let us, then, omit the conjectures of men who know not what they say, when they speak of the nature and origin of the human race. For some hold the same opinion regarding men that they hold regarding the world itself, that they have always been... They are deceived, too, by those highly mendacious documents, which profess to give the history of many thousand years, though, reckoning by the sacred writings, we find that not 6,000 years have yet passed.[26]

It is true that Augustine's beliefs were different in places to those of modern creationists. He seems to have been influenced by the Jewish-Alexandrian scholar Philo in believing God had created everything *all at once*, and that the six days were symbolic of a perfect number. His position was based upon a Latin translation of a verse from the Apocrypha (Sirach 18:1). This however failed to translate the Greek adequately. It reads; 'He that liveth for ever created all things together [or simultaneously]' or in Latin *qui vivit in aeternum creavit omnia simul*[27]. This appears to have coloured Augustine's belief. However, in the earlier Greek the last two words read as *panta koinee*. This can be translated as 'all things in fellowship,' which implies that God created the world as an integrated whole.[28]

---

[25] Whitcomb and Morris, *The Genesis Flood*, pp. 474-484.

[26] Augustine, *The City of God*, Book XII: Ch 10, 'Of the falseness of the history which allots many thousand years to the world's past.'

[27] This is taken from Douay Rheims, 1899, American Edition (a Catholic translation from the Vulgate). Although Jerome used the MT to translate the main Hebrew text he referred to the LXX and Vetus Latina for parts of the Apocrypha.

[28] See for instance Zuiddam B (2010) 'Augustine: young earth creationist,' *Journal of Creation*, 24(1) 2010 pp. 5-6 (The full text of an interview of which a summary in Dutch appeared in *Reformatorisch Dagblad*, 15 April 2009). See also Galling P

Augustine's concern over the Genesis account related to the mention of morning and evening on days one to three, prior to the formation of the sun and moon on day four. Augustine speculated in *The City of God* that the light may have been material, or that it may have been the light from the heavenly city shining upon the newly formed earth. However, he urged us to believe it whether we understand it or not. He writes

> …what kind of light that was, and by what periodic movement it made evening and morning, is beyond the reach of our senses; neither can we understand how it was, and yet must unhesitatingly believe it.[29]

Old earth creationists highlight the presence of light prior to the formation of sun and moon as a major cause of their objection to a literal six days. However, it may be noted that a 24-hour day does not depend on the position of the sun, but is dependent upon the spin of the earth on its own axis in relation to a fixed point. At the poles we may note that a 24-hour day does not require the sun to set below, or rise above, the horizon.

For Augustine the earth's literal age was determined by reference to Scripture, and this was in opposition to the pagan texts that existed at that time, particularly those of the Manicheans. As discussed already, the ancient pagan nations from Africa, Europe and the Middle East had a belief that history extended back for hundred's of thousands or millions of years, or apparently billions of years for the Hindus. But for the early Christians it was held that 6,000 years had not yet passed since the foundation of the world.

---

and Mortenson, T, 'Augustine on the Days of Creation: A look at an alleged old-earth ally,' *Answers in Genesis*, January 18, 2012
[29] Augustine, *The City of God*, Book XI: Ch 7, 'Of the Nature of the First Days, Which are Said to Have Had Morning and Evening, Before There Was a Sun.' See also Augustine, The City of God, Book XI:6 & 7, "That the World and Time Had Both One Beginning, and the One Did Not Anticipate the Other" and "Of the Nature of the First Days, Which are Said to Have Had Morning and Evening, Before There Was a Sun". Terry Mortenson writing in *The Great Turning Point*, Master Books, Green Forest, AR, pp. 40–41, 2004, suggests a movement towards a more literal reading of the creation account in Augustine's writing in later life. Augustine, The City of God, Book XII: 12 & 14, "How These Persons are to Be Answered, Who Find Fault with the Creation of Man on the Score of Its Recent Date" and "Of the Creation of the Human Race in Time, and How This Was Effected Without Any New Design or Change of Purpose on God's Part".

Chapter Summary

It would seem then that amongst the Church Fathers there was a widely held belief in a recent creation during the few centuries following the events of Christ's life on earth. For Irenaeus, and several others, this was linked to a millennial scheme; the six days of Creation prefigured 6,000 years of Earth history, followed be the seventh 'day' as a millennial period of rest. Irenaeus believed this had passed down apostolically through John. Justin Martyr, Hippolytus, and Theophilus held to similar millennial schemes. Basil continued to hold to a belief in a literal creation account in the fourth century, despite Origen's influence upon Christian thought. A literal-symbolic interpretation is also present in the writing of Augustine. Augustine saw symbolism relating to the person and work of Christ throughout the Old Testament, but also believed in a recent Creation and in a global Flood. Although Augustine struggled to accept six literal days he did not believe history extended backwards for millions of years. Instead he thought God had created all at once, seemingly with influence arising from Philo and the Alexandrian school of thought. He thought that the light of the first three days might have been spiritual light from the heavenly city, and possibly not physical light.

In the writing of Basil and Augustine there was rejection of the spirit-matter dualism of Plato and the Gnostics. Real events in history spoke symbolically of Christ and the Church, and they would not have wanted to separate creation so strongly from the Creator as modern science demands.

However, one area where the Church Fathers went wrong was to accept uncritically the best Aristotelian science of their day; that is specifically geocentrism and spontaneous generation. As will be discussed later, the well-documented Galileo affair shows how Galileo effectively challenged Aristotelian geocentrism in his time. Incidentally, Lennox argues that because the Church abandoned geocentrism in light of the advance of science then so too creationists today should abandon belief in a recent creation because of modern science.[30] The problem with this is that belief in geocentrism primarily arose from pagan sources and was supported by a few poetic biblical proof texts (*Psalm 93:1*). On the other hand, the six-day account of Creation forms a central part of the biblical narrative and was not supported by the Greek thinkers. For this reason, comparisons between the Galileo affair and young earth creationism do not hold together. It is true that Basil and Augustine apparently struggled to come to terms with the biblical timeframe of 24-hour days during the

---

[30] Lennox, *Seven Days that Divide the World*, 2012 (This is one of the central arguments of his book).

Creation week, but this is because of their understanding of the activity of an eternal God acting within created time. It is however doubtful whether secular science could investigate the activity of God in forming the world. As Augustine says, we should believe it nonetheless.

Augustine's teaching on the *rationes seminales* was apparently influenced by Aristotle's belief in spontaneous generation, and arguably informed belief in the plastic theory of fossil formation that Steno challenged. It also supported continued adherence in the early modern period to the idea that life could arise spontaneously out of the ground. And in the present day, to belief that life arose by chance out of primeval soup. Steno's effective challenge to a belief in the inorganic plastic theory of the origin of fossils will be discussed later. But regrettably, a belief in spontaneous generation by chance remains in the beliefs of modern Darwinists.

4.

The Influence of Pagan Beliefs on Christian Thought

There is a need now to look at the pagan ideas that continued to influence Christian thinking, and even hindered the development of science in the early modern period. This is quite a complex story, but it is worth spending time getting to grips with it.

In the pre-modern period there was an emphasis on the symbolic reading of nature. This can be traced back to the work of early Christian thinkers who had flirted with aspects of Gnosticism with its Platonist' influence, particularly amongst those who followed the Alexandrian school of thought where matter and spirit were separated. As such the Christian mind was focussed upon the spiritual dimension of life. Christian thinkers were seeking to comprehend God through contemplation alone, to the point where physical reality was neglected and considered to be of less importance. This attitude could not lead to the development of science. There was though a recovery, through Aquinas, towards the Aristotelian view that mental thought must first be in the senses. This led to greater harmony between the mind and sensory experience. However, it did not heal the division between the spiritual and the material. If anything it tended towards atheism where the mind, through sensory experience, became focussed upon the material to the exclusion of the spiritual. Such an approach to the scientific investigation of the world then tends to lead away from transcendent principles. However, it is ultimately inadequate because science must still make important prior commitments that are not testable through sensory experience.

The Influence of Hermeticism

There was also a lot of interest in pagan writings amongst several Christian scholars in the late Middle Ages and Renaissance period that was not especially healthy. Their commitment was that an original revelation of religious truth had been passed down to the first human beings by divine agents and later safeguarded and transmitted in the writings of such authors as Hermes Trismegistus, Zoroaster, Orpheus, the Druids, Brahmins, Plato, Pythagoras and the Sybils. These writings (i.e. the works of Plato, the Hindu writings, the *Chaldean Oracles* and *Sybilline Oracles, Orphica, Hermetica* and the *Golden Verses* etc.) were preserved and treasured as legitimate sources of religious knowledge and wisdom. However, it was considered that only a select few Gentiles were entrusted with this secretive esoteric knowledge, thus establishing a rigid class system that was not conducive to good education for all in an open society.[1]

A number of Christian writers were then unhealthily influenced by the pagan belief system, although it was generally considered that the ancient theology had been given first to the Patriarchs such as Adam, Noah and Moses, and then passed to the Greeks, Egyptians and Babylonians. It was believed for instance that Moses had passed his wisdom to the alchemist Hermes Trismegistus (meaning 'thrice great') through association with Egypt. However, this Hermes is widely considered to be a mythical person and instead identified as a combination of the Egyptian god Thoth and the Greek god Hermes. Both gods were associated in some way with divination, medicine and Hades, and Hermes became identified with alchemy, astrology and the practice of magic. However, on the other hand, John Marsham, with his work *On Egyptian, Hebrew and Greek Chronology* published in 1671, argued that the flow of knowledge had been from Hermes to Moses giving pagan metaphysics the pre-eminence over Scripture. But the Hermetic writing, with its claim for immense ages, came under close scrutiny from Isaac Casaubon, who argued in 1614 that the *Hermetica* had in fact been written in the second or third century AD. Although Robert Fludd, a Renaissance philosopher, ignored Casaubon's argument, others such as Isaac Newton responded by claiming that although the pagan document may be more recent, the knowledge contained within had been passed down faithfully.

The *Hermetica* therefore also influenced Newton, although writing in *The Philosophical Origins of Gentile Theology* he also asserted that true monotheistic religion had passed down through Noah and his offspring who later imparted this belief system to Egypt. He argued though that the original monotheistic faith was heliocentric and astronomically and mathematically rational providing evidence of a great designer. But, according to Newton, Egyptian priests, as the descendents of Mizraim, later intertwined this rational belief system with esoteric symbolism and an entire pantheon. Newton had an interest in alchemy and astrology from the works of Hermes, and believed the ancient temples of the Egyptians, Zoroastrians and Brahmins evidenced this symbolic faith. These temples he believed consisted of a central fire surrounded by the mathematically precise structure that represented the heliocentric solar system, where the fire was symbolic of the sun.[2] Newton's own beliefs tended towards Arianism rejecting the deity of Christ, although he considered himself to be a Bible believing Christian. But the point of note here is that as Christians we should acknowledge that there has long been a pagan dimension to the

---

[1] McCalla, *The Creationist Debate*, p.40
[2] McCalla, *The Creationist Debate*, p.45

study of nature. We should be careful to note that a number of our heroes of science had an unhealthy interest in pagan beliefs.

Fludd's Renaissance beliefs included astrology, alchemy, Christian cabbalism and neo-Platonism. However, Fludd came to view most of the ancient texts of Greek philosophy as corrupt and diabolical being founded on the world's wisdom, and therefore sourced from the 'Prince of Darkness.'[3] But Fludd did not reject Plato, whose philosophy he believed was acceptable, because he thought that Plato had read the five books of Moses and was therefore a source of light. From the Middle Ages Christian thinkers found much attraction in Plato's work despite his affinity for the existence of a pantheistic world soul, or *anima mundi*, and rejection of creation out of nothing. Acceptance of Plato's beliefs also raised interest in other Hermitic works where the earth was considered a living thing and nature a *mother*.[4]

The Problem of Plato and Aristotle

Aquinas had learnt of Plato and Aristotle's work through the influence of Islamic scholars such as Averroes who had helped to keep the Greek beliefs alive. The Hellenistic philosophy had initially passed into Islamic culture in the ninth century through translation by Muslim scholars. They considered the works of Plato and Aristotle to be equal to the Judeo-Christian Scripture in providing inspiration for reading the Quran. This body of ancient Greek literature was then translated back into Latin in the twelfth century. Christian scholars set about organising these works and the project became increasingly rationalistic. As such it was held that God was bound by his own natural laws, and had made the cosmos perfect as the best possible world with no further intervention necessary. That is, Greek rationalism was leading to an increasingly deistic concept of God. Averroism, as it became known, later developed into Christian heresy by the late thirteenth century where theology was considered subordinate to the empirical study of nature. However, Aquinas had been keen to uphold Scripture, and was careful to place a limit on the potential achievement that human reasoning could gain. He asserted for instance that Scriptural revelation limited the age of the world, whereas Aristotle had believed the cosmos was eternal and cyclical. Averroism however became established in the University of Padua in Italy by the time Galileo entered that establishment as a student and later tutor. Galileo must therefore have been familiar with and influenced by the Platonism of Averroes' thinking with

---

[3] Harrison P., *The Bible, Protestantism and the Rise of Natural Science*, Cambridge: Cambridge University Press, 1st Paperback Ed., 2001, p. 106
[4] Harrison, *The Bible, Protestantism and the Rise of Natural Science*, pp. 40-41

its dualism between matter and spirit, and this influenced his own approach to science and faith.

One of the problems with Plato is that his form of writing, for instance the dialogue in the *Republic*, lends itself towards relativism because concepts may change their meaning as the discussion develops through the dialogue. There is then uncertainty over what is implied, and many books have been written over Plato's intentions. Leo Strauss for instance argued that there is a 'double truth' in many works of philosophy with a hidden meaning for the elite and a plain sense reading for the majority, perhaps even to the point of deliberate deception by 'noble lies.' Strauss commented that Plato's work was both an instruction to the elite to rule over the populace and maintain social order, and an imploration to the gods to keep them docile, although he notes that this distinction is often blurred in philosophy.[5] Karl Popper in *The Open Society and Its Enemies* noted that the political theory in the *Republic* was essentially totalitarian. In the first volume, *The Spell of Plato*, Popper pointed out that philosophers have been seduced through history by the apparent greatness of Plato and overlooked the inherent tyranny in the *Republic* with its deceitfulness, violence and rigid class structures. This work also seemingly promoted the break up of families, forced labour, fascism and eugenics. The fact that the *Republic* was written as a dialogue allowed Plato to leave ambiguity over how it should be interpreted, but Popper accused Plato of betraying his teacher Socrates. According to Popper, Plato was driven by a fear of change that an open society would produce.

The ideal society for Plato was comparable to the soul of man and divided accordingly into appetite (abdomen), spirit (chest) and reason (mind). The productive side of the soul corresponded to the abdomen, and in the ideal society consisted of lower class workers and labourers. The chest part of the soul represented the adventurous spirit where the warrior class was expected to act as the protectors of society, while the governing class, as philosopher kings represented the head or reason of the body. Plato considered it necessary to force this class division through education with family ties broken. Children were to be nurtured by the state and monogamy was undermined. Does this resonate with our modern broken society?[6]

---

[5] See for instance: Fuller, S, *Science vs Religion?*, Oxford: Polity Press, 2007, pp. 52-52; commenting on, Strauss, L. *Persecution and the art of writing,* Chicago: Chicago University Press, 1952

[6] There is incidentally some commonality between Plato's *Republic* and the Hindu caste system where high-class priests are honoured in power and surrounded by a high caste military, while many others are in subservient classes, and millions of

However, it is of interest to consider how far the ancient pagan theology extended into the later scientific age. Any secular account of the development of science will give the impression that modern science is entirely empirical and has nothing to do with such dualistic considerations. The perceived battle between science and religion has been cast as one between the Judeo-Christian Scriptures and empirical evidence, but there are important considerations relating to the influence that other pagan religious beliefs have had upon science. It may be noted further that there has been a tension through history between the Bible and ancient Gentile pagan theological texts. For much of Church history Christianity accommodated these texts by seemingly Christianising them, but with the Reformation came a greater commitment to the literal text of the Bible with the symbolic Renaissance interpretations downplayed. The Reformation, and the translation and printing of Bibles, enabled the corresponding scientific reformation in large part. Science developed because it opened up a way of reading the created order literally, and encouraged respect and equality in education and the pursuit of knowledge. As Peter Harrison has noted, a more literal reading of Scripture encouraged a more literal reading of nature, which led to developments in operational science.[7] However, aspects of the ancient pagan theology were carried over into the modern period as well with astrology becoming astronomy, and alchemy developing into chemistry.

Chapter Summary

Although the Platonists and Pythagoreans had a great deal of knowledge in antiquity, and undoubtedly many proponents were very bright, the rigid class system epitomised in Plato's *Republic* severely hindered the development of science in the ancient world. This is because it severely restricted access to knowledge and individual achievement. Plato's work continues to influence science and society to this day, where there is evidence of an ongoing dualism between the material and the spiritual in science, for instance through concepts such as methodological naturalism. It is also evident that leading scientists tend to be idolised to the level of philosopher kings, this reinforced today by institutions such as the Royal Society. However, perhaps ironically, it may be noted that some early members of that society held deeply religious beliefs. But there is a case to be made for a different, Christian approach to science that rejects the strict dualism between the material and spiritual, and one that instead develops a

---

people are subjugated to the level of Dalits or untouchables who are even placed outside of the class system.

[7] Harrison, *The Bible, Protestantism and the Rise of Natural Science*, 2001

34

commitment to first class education for all with the rigid class structures removed. Michael Polanyi for instance has commented on the importance of only general authority in science based on freedom, truth and conscience.[8] The following discussion through later chapters will be mainly focused on questions relating to the re-emergence of belief in long ages and the development of geology in the modern period and how that feeds into evolutionary thinking.

---

[8] Polanyi, M. *Science, Faith and Society*, London: Oxford University Press, 1946, pp. 47-49

5.

Copernicus, Galileo, Kepler and Bacon

The latter part of the sixteenth century and the start of the seventeenth century saw a transformation in the way the world was viewed, from the pre-modern Renaissance thinking to the early modern period involving the development of science. The study of the natural world then moved away from a symbolic interpretation of nature to a more literal one; one that could be experienced through the senses and understood rationally in the mind. This developing science was conceived partly as a means of recovering knowledge that was thought lost due to human sin. The influence of Augustine led to the belief that sin had dulled the human mental capacity (the noetic effect of sin), and knowledge gained through the senses was therefore considered unreliable. Francis Bacon argued that in order to recapture the knowledge that Adam had lost in the Garden of Eden, it was necessary to study nature in-depth through the development of a rigorous process of experimentation. So, whereas Adam's perfect knowledge was given by grace, Bacon believed that man must now work with the sweat of his brow to recover that same knowledge.[1]

The scientific approach was then developed as a response to the pre-modern philosophy that tended to focus upon the symbolic reading of nature with less concern for the material. Reading the book of nature had been about recognising the intrinsic signs in material objects and living creature, so that their natural true essences might reveal something of the divine mind. Even though mankind had fallen from grace it was believed by various philosophers, such as Paracelsus, that it was still possible to read the signs because of a divinely given light of nature. From this Paracelsus was prescribing medicine on the basis of these signs, not according to their empirically tested usefulness. However, the scientific revolution brought the study of nature into a realistic perspective where nature was studied literally in order to discern the mind of God. The Protestant Reformation was influential in the scientific revolution where a more literal interpretation of Scripture, and greater respect for the plain sense text, allowed nature to be studied more literally, as Peter Harrison for instance has recently noted.[2]

---

[1] Whether Bacon was really trying to gain lost knowledge, or access of forbidden knowledge is an important question to consider, but beyond the scope of this study
[2] See for instance; Harrison, P, *The Bible, Protestantism and the Rise of Natural Science*, Oxford: Oxford University Press, 2001

The Rejection of Geocentrism

But there continued to be mystical elements of Renaissance thought in certain Roman Catholic circles in the sixteenth and seventeenth centuries. A number of scholars held that fossils grew within the ground because of some esoteric plastic force or lapidifying power (*succus lapidificatus*) acting on inorganic material. Albert of Saxony espoused this idea in the fourteenth century, developing it from Avicenna's Aristotelian beliefs of the eleventh century. Rudwick comments that it was widely held by Catholic philosophers even towards the end of the sixteenth century.[3] The Ancient Greeks however showed less interest in the significance of fossils, and where they were discussed in any meaningful sense it was in terms of the remains of legendary creatures or heroes such as the Cyclopes.[4] This esoteric belief, involving plastic forces at work in nature, was also in harmony with the alchemists and the Hermitic tradition.

But it wasn't until the time of the Thirteenth Council of Trent in 1561 that Aristotle's philosophy was accepted more formerly into the Catholic Church, and it was this that ultimately led to the controversy over Galileo's work. One view is that this came about partly because the Council had developed a definitive version of the Eucharist known as transubstantiation. This doctrine was based on the natural philosophy of Aquinas and Aristotle with its distinct theory of motion. However, following the outcome of this Council, Father Aquaviva, the Superior General of the Jesuit order at the end of the sixteenth century, commented that professors should not deviate from Aristotle in matters of important philosophy.[5] As a result Aristotelian-Ptolemaic geocentrism, the idea that the earth was fixed in the heavens with the sun moving around the world, became an important part of Catholic dogma. The plastic theory of fossil formation was also favoured because of the influence of Aristotle at this time, as will be discussed more fully in later chapters.

But Nicholas Copernicus, a Canon in the Catholic Church, found increasing necessity for epicycles (cycles upon cycles) to account for the motion observed in the planetary systems as a result of the hypothesis of geocentrism. This he considered inadequate and he sought a more simple solution. However, we can also see the influence of Plato, and his metaphor of the sun, in the thinking of Copernicus with the universe and sun

---

[3] Rudwick, M. J. S. *The Meaning of Fossils: Episodes in the History of Palaeontology*, University of Chicago Press, 1985, p. 24
[4] Sedley, D. *Creationism and its Critics in Antiquity*, pp. 43-44
[5] Blackwell, R.J., *Galileo, Bellarmine and the Bible*, Notre Dame, 1991, pp. 122, 141

described according to its most beautiful form as a temple (*pulcherrimo templo lampadem*). Copernicus wrote further that

> It may be aptly known as the lantern of the world (*lucernam mundi*), its mind (*mentem*), or by others its ruler (*rectorem*). Hermes Trismegistus calls it the visible god (*uisibilem Deum*), while Sophecles Electra the all seeing (*intuente omnia*).[6]

The Platonic theme of beautiful forms is also present in the writing of Kepler, Newton, and later even in Charles Darwin's *Origins*. Astronomers, such as the sixteenth century Danish Lutheran Tycho Brahe, were able to map the sky with increasing precision and this enabled the development of astronomy as a science. As a result of the disparity between theory and observation Copernicus had proposed his heliocentric theory in 1543 because the mathematics fitted the evidence better. One notable supporter who embraced the new idea of Copernicus was Michael Maestlin of the Tübingen Lutheran University in Germany. However, the new theory failed to gain widespread appeal, initially because it seemed from mere common sense that the earth was fixed in space. A moving sphere they believed would cause great turbulence with objects and people thrown off.[7]

It was a student and acquaintance of Maestlin at the Tübingen University, Johannes Kepler, who further developed the new theory of Copernicus. Kepler had taken an interest in astronomy from observing a lunar eclipse and comet as a child, and at the age of 22 in 1594 became a lecturer in mathematics at the Seminary at Graz. It was here that Kepler developed his ideas in astronomy believing that God formed the universe according to elegant and perfect mathematical regularities. This was published as *Copernican Cosmographic Mystery* in 1595 with the approval of the Tübingen University Senate. A few years later in 1600 Kepler moved to Prague to work with the Lutheran Brahe who was employed as mathematician to Rudolph, the Catholic Holy Roman Emperor.[8] Initially Kepler and Tycho were rivals, but through correspondence gained a mutual respect that led to brief cooperation, before Tycho died in 1601. Tycho never fully accepted Copernicus' theory, seeking instead to blend

---

[6] Copernicus, N., Book 1, Chapter 10 (my translation with Latin in brackets). See also, Introduction to *De Revolutionibus* 1543 trans. By Rosen 1978, p.7, 22. This was pointed out by Snobelen, S., 'This most beautiful system': Isaac Newton and the Design Argument', in, God, Nature and Design Conference, Ian Ramsey Centre, Oxford, 10-13 July 2008.

[7] Hoskin, M. *Cambridge Illustrated History of Astronomy*, Cambridge University Press, 1997, p. 32

[8] Marston, P., 'Johannes Kepler,' *CIS Online Magazine*, No.4, Autumn 2007

Copernicus' work with Ptolemaic cosmology; with both the earth and the sun fixed in space. But his skill at measuring the orbits of the planets with great precision was of great benefit to Kepler, and to his research into the mathematical regularities of the heavens. Like Tycho, the Lutheran Kepler continued to work under Catholic Emperors in subsequent years. In the following two decades Kepler developed his ideas with his first two laws of planetary motion written up in 1609 in the *New Astronomy,* and the third in 1619 as the *Harmony of the Universe.*[9] In all of this Kepler saw mathematical elegance and beauty as being a natural part of God's good creation.

Kepler was also evidently reworking Galileo's thoughts, publishing his ideas on these matters in 1611 as *Dioptrice* and developing Galileo's telescope. Although a devout Lutheran, Kepler also had an interest in astrology, and some of his ideas were influenced by Plato, seeking for instance to fit the orbits of the planets inside regular Platonic shapes of increasing sides; the orbit of Mercury for instance fitting within a triangle around the sun. With echoes of Hermetic and Egyptian influence, Kepler also came to view the sun as somehow representative of God as the Father, the orbit of the earth as the Holy Spirit and the earth representative of Christ. Kepler's consideration of theology in light of neo-Platonism in this way left him open to attack by Galileo about the extent of his knowledge of pantheism. However, Kepler's best work was developed from observations of the movement of planets. Theory was developed from those observations, once he had abandoned the Platonic scheme in favour of one that allowed all mathematical functions to inform his work.

The Catholic Church moved against the ideas of Copernicus, just as Kepler and Galileo were gaining success in their development. As a result the Sacred Congregation of the 5th March 1615 suspended Copernicus' work *De Revoltionibus orbium coelestium* until suitable corrections could be included. The assertion of the Congregation being to prevent an opinion considered ruinous to Catholic truth from gaining wider acceptance. Carmelite Father Foscarini had supported Copernicus in a letter dated 6 January 1615 adding fuel to the disagreement. The head of the Sacred Congregation, the Jesuit Bellarmine also cautioned Galileo not to defend the Copernican theory in terms of reality, although a mathematical symbolic approach was considered acceptable. In a letter to Foscarini, Bellarmine stated that he wanted proof of the Copernican theory before he would abandon his Aristotelian interpretation of Scripture that the earth

---

[9] Kepler's three laws are as follows. 1. The planets orbit with the sun at one focus of an ellipse. 2. The planets sweep out equal areas in equal times. 3. The squares of the periods of the planetary orbits are proportional to the cubes of their mean distances from the sun.

was fixed and the sun moved.[10] However, Galileo became convinced of Copernicus' theory that the earth moved around the sun from his own study of the phases of venus, and from plotting the progress of tides on the earth. Therefore, he must have found it difficult keeping quiet under the authority of Rome when the Protestant Kepler was able to study and publish the same ideas under the protection of Catholic masters. Galileo was in fact critical of Kepler's theory of the tides accusing Kepler of ascribing 'occult properties' to nature because Kepler argued that the moon had the greater power over the waters. It was Kepler who was later proved right in terms of the science. However, it wasn't until 1632, following the death of Bellarmine, and after his friend Cardinal Barberini had risen to the Papal seat, that Galileo finally published his view of the Copernican theory in *Dialogue on the Two World Systems.*

For this Galileo was condemned, placed under house arrest and forced to recant, mainly for resisting Church authority and seemingly abusing the Pope. Paul Marston argues that Galileo was not as great a scientist as Kepler and only joined the debate because of a love of argumentation, but the competitive nature of this lively discussion led Galileo to improve his scientific work even as his tact failed him. In Galileo's *Dialogues*, the argument of the Pope, who called for avoidance of dogmatism on scientific matters, was put into the mouth of the dunce. Perhaps Galileo conceived this as a criticism of the previous pontiff, but it caused great offence nonetheless to his former friend and supporter.[11] The dispute between geocentrism and heliocentrism however may also be seen, in part, as a dispute between the legacy of the works of Aristotle and Plato.

Galileo was aware of the Augustinian tradition of interpreting Scripture, and sought to defend his work within that tradition, at least as far as it suited him. He set out his views on the relationship between truth in science and truth in Christian faith in a letter to Grand Duchess Christina of Lorraine in 1615. In this he explained the manner in which it is possible to hold to the integrity of Scripture and accept the new science. Firstly, according to Galileo, the book of nature and the book of Scripture cannot contradict one another, both seek after truth and they are always complimentary as they proceed from God, although sometimes speaking in different arenas. Secondly, science can and should provide a path to truth that is independent of Scripture, and thirdly, Scripture cannot be used to deny scientific truths established through experimentation and observation. He was arguing then that if a clear and pressing observation in nature conflicted with a literal reading of Scripture, then the literal meaning of that

---

[10] Cited in Forster, and Marston, *Reason, Science and Faith*, 1999, p.319

[11] Sharratt, M. *Galileo: Decisive Innovator*, Cambridge: Cambridge University Press, 1994, pp. 169, 175

understand. Calvin, for instance writing in *Commentaries on the First Book of Moses called Genesis*, commented that the intention of Moses was to write in a popular style so that ordinary people may understand it.[22] McCalla comments that the Protestant Reformers were willing to accommodate Scripture in this way in light of science. However, we may note that both Calvin and Augustine held to a recent creation and held Scripture to be the more reliable source of truth. Galileo though had gone a step further than Calvin and effectively subordinated the book of Scripture to that that of the book of nature.[23] For both Kepler and Galileo mathematics was the language necessary to decode the complexity found in nature.

Francis Bacon and the Scientific Experiment

For Bacon the process of science was to be carried out through empirical experimentation using inductive reasoning under the guidance of a scientific judiciary. He was also working in response to the symbolic Renaissance interpretation of the book of nature. He considered the attempt to read the symbols of nature by Renaissance theologians as impossible and arrogant because Adam's Fall had corrupted human minds. Just as Adam had been thrust from the Garden of Eden and then had to work by the sweat of his brow, now the natural philosopher had to study nature through rigorous and careful experimentation. Only through such a process could the researcher escape the deceitful effects of the fallen mind and recover the lost knowledge. Bacon called this new process his new organ or *Novum Organum* and the quest to recover that knowledge involving scientific courts and crucial experiments (*experimentum cruces*). He saw this as a *Great Instauration*. Bacon was for instance concerned to counter an apparent abuse of Scripture that he believed evident in the Renaissance group known as the school of Paracelsus. As noted this group of scholars were suggesting that all science should be done through symbolic interpretation of Scripture whilst ignoring observations in natural philosophy. Bacon commented that.

> The school of Paracelsus, and some others...have pretended to find the truth of all natural philosophy in the Scripture; scandalising and traducing all other philosophy as heathenish and profane. But there is no such enmity between God's word and his works. Neither do they give honour to the Scriptures as they suppose, but embase them. For to seek heaven and earth in the word of God, whereof it

[22] This is cited in McCalla, A, *The Creationist Debate*, p. 5
[23] McCalla, *The Creationist Debate*, p. 7

is said 'heaven and earth shall pass away but my word shall never pass away' is to seek temporary things amongst eternal; and as to seek divinity in philosophy is to seek the dead amongst the living...And again, the scope or purpose of the Spirit of God is not to express matters of nature in the Scripture, otherwise than in passage, and for application to man's capacity and to matters moral or divine.... In this vanity some of the moderns have an extreme levity indulge so far as to attempt to found a system of natural philosophy on the first chapter of Genesis, on the book of Job, and other parts of sacred writings; and repression of it is the more important, because from this unwholesome mixture of things human and divine there arises not only fantastic philosophy but also an heretical religion.[24]

Bacon here asserted that the purpose of Scripture was not to provide a basis for all natural philosophy, but to provide a proper understanding of morals and the divine character, and Bacon noted that science, and Scripture need not be in conflict. Therefore Bacon was not seeking to claim that Scripture can say nothing about science, but to counteract the inappropriate use of Scripture by the school of Paracelsus where proponents were seeking to base *all* natural philosophy on a symbolic reading of God's revealed word. However, Bacon's writing's have been used subsequently to argue that Scripture has *nothing* to say with regard to science. In other words, the commonly held view of the Baconian approach to the science-faith dialogue is not entirely correct. It may be noted further that Bacon mentions that the Scriptures may express matters of nature 'in passage' which may in itself inform our study of nature. Mortenson also notes that later scriptural geologists, such as Granville Penn, considered that they were working within Baconian criteria, where the study of Scripture was allowed to speak as an accurate historical record, but that Scripture and nature should not be unwisely mixed.[25]

Bacon went further and argued that through his new methodology mankind could recover the true knowledge of nature that had been lost as a result of Adam's Fall in the Garden of Eden. Bacon's approach was not totally devoid of grace commenting on the necessity for divine revelation. He wrote

---

[24] Bacon, F., *Advancement of Learning*, Oxford Ed, Book II, Part XXV.16, 1906 (1605), p. 229
[25] Mortenson, T. *British scriptural geologists in the first half of the nineteenth century- part 1: Historical setting.* TJ (JoC), 11(2), 1997, pp. 221-252. See also Mortenson, T. *The Great Turning Point,* Grand Rapids: Master Books

For God forbid that we should give out a dream of our own imagination for a pattern of the world; rather may he graciously grant to us to write an apocalypse or true vision of the footsteps of the Creator imprinted on his creatures.[26]

As the King's Lord Chancellor and chief lawyer, Bacon approached science with a legal mind and proposed the idea of a crucial experiment to judge competing knowledge claims. Accordingly, proponents must set out their claims humbly so that an independent judge can lay the ground rules for exploration and then decide between the hypotheses. Although Bacon's concept of a scientific judiciary seemingly failed to develop, later the Royal Society took up the role forming a collective body of opinion.[27]

Whilst at face value Bacon's proposals seem perfectly compatible with Christian ideals, the quest to recover that lost Edenic knowledge through experimentation can easily be corrupted and turned into a quest for forbidden knowledge, the fruit from the *tree of the knowledge of good and evil*, and not simply a recovery of lost knowledge. The search for hidden knowledge was also the desire of the alchemists and claims persisted that Bacon and the later Royal Society were motivated by unholy desires. The idea of an elite scientific judiciary may also be seen as close to Plato's political philosophy and social order. In the *Republic* Plato believed that philosopher kings should over rule the ideal city-state *Polis*, with the help of a high class military. This perhaps still permeates the mindset of the Royal Society that today seeks to be the gatekeepers of scientific knowledge. This is epitomised for instance by the dismissal of Michael Reiss for daring to suggest creationist beliefs should at least be treated with respect in the classroom.

For these (and other) reasons there have been suggestions that Bacon was in fact Rosicrucian in theology, and involved in helping to establish modern Freemasonry. He was for instance given the role of spiritual advisor to William Shaw. King James I had tasked Shaw with organising Scottish Freemasonry into the Scottish rite. Bacon also wrote the *New Atlantis* in the early years of the 1600s, in which a proposal to settle America has been suggested by some, but it may be an appeal to establish a society on the beliefs of the legendary people of Atlantis. In Plato's *Timaeus* and *Critias* Solon, the founder of Greek democracy, had

---

[26] Bacon, F., 'Great Instauration: Plan of Work,' [1620], translated into English by Ellis, R., in Spedding, J., *Works of Francis Bacon*, 1858, Part VI. Cited in McCalla, *The Creationist Debate*, p. 9.
[27] Fuller, S., *Science and Religion? Intelligent Design and the problem of Evolution*, Polity Press, 2007, pp. 63-66

learnt about the lost island of Atlantis from Egyptian priests. Atlantis it would seem from legend had been lost to a catastrophic Flood, and the philosophy of the ancient inhabitants was that of the Egyptian religious order. It was perhaps upon the beliefs of the mythical people of Atlantis then that Bacon wanted to establish his new society. Baigent and Leigh comment that prior to the establishment of the Royal Society, a group of scientists and esoteric philosophers had formed an 'Invisible College' that was really Rosicrucian in nature. During the English Civil War this Invisible College included such notable characters as John Locke and Robert Boyle, but it was not until 1660 that the members of the Invisible College finally established the Royal Society openly.[28]

While Bacon's proposition was for a new society to develop science through crucial experimentation, later members of the Royal Society continued to look for the signatures that they believed had been placed in nature by God, but lost due to the effect of Fall. On top of this the original language that Adam and Eve had spoken in the Garden of Eden had become corrupt, or lost as a result of a second fall at Babel. Discussion ensued over whether the original language was Hebrew, Egyptian or Chinese, and the Royal Society for instance commissioned Robert Hooke to investigate the Chinese language as the prime candidate.[29] The quest for a universal language was considered necessary to overcome the confusion of languages, and formed part of a greater quest to recover the pre-Fall knowledge that Adam once possessed. Harrison comments that this only led to further allegations that the Royal Society was really serving a Rosicrucian agenda using a form of cabbalism to recover that forbidden knowledge.[30]

McCalla makes similar claims relating to the Jesuits. In the early modern period there continued to be concern that the ancient Gentile pagan theology, with its symbolic interpretation of nature, was being carried over into the emerging science and that suspicion fell most heavily on the China Jesuits, the Italian Jesuit Athanasius Kircher, and Robert Fludd. The Jesuits had travelled widely and many were missionaries in China by the end of the sixteenth century. As part of their plan for evangelism they felt it necessary to adapt the chronology of the Bible to fit with Chinese culture and history. The length of the Chinese Dynasties meant that they wanted to adopt and accommodate the Septuagint with its longer Biblical history that better fitted with Chinese history.[31] However, there also seems to have been an

---

[28] Baigent, M. & Leigh, R, *The Temple and the Lodge*, Jonathan Cape, 1989, p. 145

[29] Harrison, *The Bible, Protestantism and the Rise of Natural Science,* p.259

[30] Harrison, *The Bible, Protestantism and the Rise of Natural Science,* pp.260-261

[31] McCalla, *The Creationist Debate*, pp. 40-42

affinity for the works of the Hindus, Egyptians, Plato and Aristotle within sections of the Jesuit movement that continued into the scientific age.

While it is possible to see in Bacon's writing a degree of ambiguity regarding his motivation, it is interesting to look in detail at Bacon's stated attitude to Scripture and how it relates to the study of the book of nature. Mortenson comments that Bacon's actual beliefs were at variance with the modern scientific understanding, and that secular scientists, in order to deny the validity of the Mosaic account of Creation and the Flood, have wrongly used the writing's of Bacon. It may be seen that modern secular views then rely on an approach to Bacon's writing that really takes it out of context. Bacon for instance commented in the *Advancement of Learning* (1605), that the study of nature would lead to meditation upon the power of God, which is evident from the works of nature.

> For our Saviour saith, 'You err, not knowing the Scriptures, nor the power of God'; laying before us two books or volumes to study, if we will be secured from error; first the Scriptures, revealing the will of God, and then the creatures expressing his power; whereof the latter is a key unto the former: not only opening our understanding to conceive the true sense of the Scriptures, by the general notions of reason and rules of speech; but chiefly opening our belief, in drawing us into a due meditation of the omnipotency [*sic*] of God, which is chiefly signed and engraven upon his works.[32]

Bacon could see God's handiwork in creation, and for him this provided information about the power of God. Bearing these comments in mind, Bacon seemed to have had a high regard for Scripture with the natural order merely throwing further light on human knowledge of God's power and wisdom. Bacon further elevated Scripture above natural philosophy, considering the latter temporary and passing and the former divine and eternal. Bacon also considered the account in Genesis to be a literal historical account of six day's duration.

> It is so then, that in the work of the creation we see a double emanation of virtue from God; the one referring more properly to power, the other to wisdom; the one expressed in making the subsistence of the matter, and the other in disposing the beauty of the form. This being supposed, it is to be observed that for anything which appeareth in the history of the creation, the confused mass

---

[32] Bacon, *Advancement of Learning*, Book I, Part VI.16, 1906 (1605), p. 46

and matter of heaven and earth was made in a moment; and the order and disposition of that chaos or mass was the work of six days...[33]

It may be noted that Bacon considered the Genesis events to have been supernatural and a reflection of the power and wisdom of God. In this sense then the supernatural works are to be considered outside the realm of natural philosophy, and therefore really part of theology. It is noteworthy though that Bacon spoke of an instantaneous creation of matter and energy, but a period of six days to bring that chaos to order, and perhaps this dual approach to creation may in fact be seen as being influenced in part by Philo and Augustine. However, Bacon further comments that Scripture may be found to be full of lessons in natural theology, for instance in the book of Job.

So in this and very many other places in that law, [of Moses] there is to be found, besides the theological sense, much aspersion of philosophy. So likewise in that excellent book of Job, if it be revolved with diligence, it will be found pregnant and swelling with natural philosophy; as for example cosmography and the roundness of the earth; *(Job 26:7)*...[34]

Bacon thus considered that it is legitimate to learn in part natural philosophy, or early science, from the Bible. As mentioned already, he was writing in response to Renaissance symbolic approaches to the book of nature, although McCalla argues that Kepler and Galileo's method of mathematical study and experimentation should be seen as being closer to the ideals of the scientific enterprise than Bacon's proposed inductive methodology.[35] McCalla notes the shift in thinking through this period was in effect a shift within, not against biblical culture; from Renaissance thought where the Bible provided hidden symbolic meanings of nature, to one where the study of nature could throw light upon the cosmological passages of the Bible. While this may be so, it is only right to note the continued influence of Greek and pagan philosophy within the development of science. The influence of Plato and Aristotle can be seen in Renaissance interpretations of Scripture, and some theologians and scientists carried this over into the emerging scientific age. In later years Newton, with his interest in the hermetic writing, also helped to continue the study of alchemy in the scientific age, and his esoteric writing on alchemy and

---

[33] Bacon, *Advancement of Learning*, Book I, Part VI.16, 1906 (1605), pp. 40-41

[34] Bacon, *Advancement of Learning*, Book I, Part VI.16, 1906 (1605), pp. 43-44

[35] McCalla, *The creationist Debate*, p. 9

hidden codes in the Bible consisted of more than a million words in manuscripts.[36]

As mentioned already, in terms of Protestant views of the relationship between science and faith, Harrison has suggested that the Reformation was essential for providing the conditions that led to the growth of western science. By promoting Scripture as the sole basis for authority within Christianity, the reformers encouraged a literal interpretation, and downplayed the metaphorical view. This allowed a new way of classifying creation based on the elevation of the written words of Scripture, and this more literal treatment of Genesis, insisted upon by the reformers, led in the seventeenth century to developments in experimental science.[37] Grinnell further argues that it was the Protestant Reformation that led to the development of the science of geology,[38] and later Protestant theologians such as Martin Luther, John Calvin and John Wesley considered the Creation account and the Flood to be literal.[39]

Certainly Protestants played their part in the gradual re-awakening and re-ordering of the study of the earth, also helping to overthrow some dualistic ideas left over from ancient Greek philosophy. These had become embedded within Catholic thought and belief. Many in the medieval Catholic Church believed that the study of the earth was really beneath the calling of the Christian, and preferred to study the higher sciences of astronomy, geometry and mathematics. Following the Greek thinkers, Christian natural philosophers considered that such exalted mathematical sciences better reflected the wisdom, goodness and mind of God. On the other hand the earth was considered to be the place of abode of the devil, and as such the study of the earth was considered a corruption of Christian faith. However, for some involved in the newly emerging Protestantism, the study of the earth provided an opportunity to demonstrate that the biblical account of a great Flood was reliable. As such it was a commitment to the accuracy and trustworthiness of God's word that led to developments in the science of geology. The account of the Flood also brought to people's attention God's sovereignty, fear of judgment and punishment for sin that helped to maintain church authority. Allied to this was a move towards a realist approach to science instead of the instrumental view that

---

[36] Linden, S.J. *The Alchemy Reader: From Hermes Trismegistus to Isaac Newton* New York: Cambridge University Press, 2003, p. 243

[37] Harrison, P., *The Bible and the Emergence of Modern Science*, Christians in Science, Public Lecture, Cambridge University, 24th May 2005.

[38] Grinnell, G. 'A Probe Into The Origin of the 1832 Gestalt Shift in Geology,' *Kronos: A Journal of Interdisciplinary Synthesis* (Kronos Press, Glassboro, N. J.) 1(4): pp. 68-76 (Winter 1976)

[39] Grinnell, 'A Probe Into The Origin of the 1832 Gestalt Shift in Geology,' 1976

some within the Roman Catholic Church had developed through the Renaissance. The Protestant Reformation provided the groundwork for science to develop, and it overcame the excessive control that had accumulated in the Roman Church. But it did not itself fully heal the division between the material and the spiritual. The senses became focussed upon matter, which led increasingly to the exclusion of the spiritual aspect of life through the Enlightenment. Although the Reformation weakened the Catholic Church's authority, a new elite of scientists arose within the Royal Society and subsequently tried to dictate the direction of scientific discovery. As a result, later Scriptural geologists and sceptics of macro-evolution were often isolated and treated with disdain. However, a literal or historical interpretation of Scripture has provided an important heuristic to science, that is one that leads to discovery, and this includes geology as will be discussed further in subsequent chapters.[40]

Protestants were however also keen to show that God's handiwork could be seen in nature, and natural theology developed along increasingly rational lines having previously been expressed in more artistic, symbolic terms. Natural theology was becoming more scientific, seeking to unravel the mind of God through the study of nature, and this was believed to reflect God's goodness. However, the evidence for the goodness of God as seen in nature was over-emphasised in the minds of some scientists with evidence for suffering, the Fall, and the chaos of the Flood downplayed. Whereas Genesis records that the earth was in a fallen state because of man's sin, and the deluge brought about because of increasing wickedness, the continued undercurrent of neo-Platonist thinking led to the belief that evil was merely caused by the absence of goodness, or perhaps God was not omnipotent. Natural theology was in this regard partly a hindrance to geology because of its emphasis on a simplistic view of God's goodness and design plan. This evidence of a design plan was believed by natural theologians to be observed in the perfect mechanistic order of nature that could be explained in terms of mathematical regularities. The idea of a cataclysmic Flood and destruction of the earth did not reflect the desire for goodness and order in the minds of many through the Enlightenment period. The idea that geological processes exhibit machine like regularity and characteristics however later influenced the deist James Hutton.

Chapter Summary

In this chapter consideration has been given to the development of more literal and scientific modes of thought, and away from the symbolic

---

[40] As noted even by Rudwick, M.J.S., *World Before Adam: The Reconstruction of Geo-History in the Age of Reform*, University of Chicago Press, 2008. p. 564

reading of nature from the pre-modern period. During this time Aristotelian geocentrism gave way to a Hermetic or Platonic sun-centred universe, although much of this was due to advances in observational science. For instance the invention of the telescope led to important new discoveries, and Kepler's best work came when he allowed all mathematical functions to inform his thoughts. The pre-modern Platonism tended to be elitist, and had primarily focused upon the spiritual aspect of thought, which did not encourage the study of nature. The Protestant Reformation led to developments in science because of its commitment to a literal reading of Scripture, and a more egalitarian approach to education.

But even though pre-modern thinking declined as the mind became focussed upon the material world instead of the spiritual, the dualism did not disappear because of a preference to do science apart from revealed faith. This dualism is seen for instance in the thinking of Averroes and Galileo. Galileo did not want to subordinate operational science to Scripture, and this independence view led eventually to the elevation of science above revealed faith. However, where Scripture is silent about the details of historical events or observations, it may still provide a general authority concerning claims about what is true. The dualism does though persist today through for instance the insistence of methodological naturalism by some philosophers of science, such as Michael Ruse. But as will be shown in the next chapter, allowing Scripture to speak about matters of science has in fact given rise to new discoveries, even for instance in the field of geology. The other Aristotelian idea that was undermined by the advance of science was spontaneous generation, firstly relating to the question of the origin of fossils, and later with organic life through the work of Pasteur. Although regrettably a belief in spontaneous generation remains embedded in the modern narrative of evolutionary science.

6.

Steno, Fossils and the Flood

One notable scientist from the early modern period was the Danish anatomist and geologist Niels Stensen; he later settled in Italy and adopted the Latinised form of his name Nicolai Stenosis, or Nicolaus Steno. Travelling from his native Denmark to further his research he first passed through Amsterdam to Leiden where he was able to give public demonstrations in anatomy. This increased his exposure to leading scientists and philosophers, including Baruch Spinoza who occasionally attended his lectures in Holland. Spinoza was keen to discuss Descartes' philosophy with Steno, and Steno was at first enraptured by these new ideas, although Steno lost confidence in the thinking of Descartes.

Descartes was considered the French equal of Bacon. For instance he published his work *Discourse on Method* in 1637 in which he argued that all science and philosophy must begin with doubt. In this there was reliance upon observation and reason as a means of establishing truth. However, Descartes was fallible and made a mistake on the function of the heart; he believed it was a generator of heat and not a pump. But Steno's own examination showed that the heart was a muscular rhythmic pump, with regular contractions of muscle fibres enabling the blood to flow. Steno also disagreed with Descartes on the formation of teardrops. As a result of such errors Steno's faith in the philosophy of the rationalist Descartes diminished. He also had a craving for certainty in matters of faith, although he lived in a time of uncertainty with people confused and unsettled by the ongoing struggle between Protestantism and Catholicism, especially over the nature of religious authority. The Catholic authorities had argued that only the clergy should interpret Scripture because the laity would be confused by difficult passages. Many Protestants were however more open to an egalitarian approach to the reading of the Bible. Steno grew up in this period and his student journal, aptly named *Chaos*, reflects this political turbulence.[1]

Nicolaus Steno and the Organic Origin of Fossils

Ole Borch, one of Steno's tutors, had an involvement in alchemy and tried to gain the interest of Steno, but Steno's interest in the subject was limited to watching solids precipitate out of water as a physical process. As noted, the Catholic Church had embraced and Christianised

---

[1] Cutler, A., *The Sea Shell on the Mountaintop*, London: William Heinemann, 2003

some of the sciences of Aristotle and Plato. But many of these ancient Greek ideas were being questioned as the Renaissance symbolic interpretation of nature gradually gave way to the emerging scientific age. Alchemy for instance was to graduate into chemistry, and astrology became astronomy. While Protestant reformers taught that ultimate truth could only be found in the pages of the Bible there were also still elements of Greek thought in Protestant circles as well. Steno seemed however to show little interest in the pagan ideas, seeking instead to develop the sciences he was working on along rational lines. For Steno however this rationality included an approach that allowed Scripture to influence his geological studies, and this was to prove fruitful.

Steno passed through Paris, but later settled in Florence with financial support from the Medici family. He became a member of the short-lived, but notable science community, the *Accademia del Cimento*. Such was his contribution to science and faith that following his death the people of Florence campaigned to have his remains brought back to the city and gave him a burial place within the Basilica di San Lorenzo in Florence. While living in Florence he continued to study anatomy, but this was not his only area of research as the geological strata and structures of the Italian landscape also captivated his interest. Although he started out as a Protestant, and was throughout his life keen to harmonise the Bible with observations in the rock layers, his interest in the science of geology withered following his conversion to Catholicism. He was then rapidly elevated to Bishop. In Italy Steno found the spiritual passion and certainty he desired within the Roman Catholic Church.

His practical studies of geology however led him to conclude that the sedimentary rock layers were laid down as suspensions in water, and that this sediment formed around the fossil. Clearly fossils could not form naturally in the rock if they were of organic origin. Steno was for instance given the head of a shark to dissect by fishermen from Livorno, and from this he noted the similarity between the shark's teeth and fossil 'tongue stones' or *glossopetrae* known from the fossil layers. These fossil teeth were seemingly identical in form to the teeth that he knew from anatomical studies. This led to the inference that they once belonged to real sharks, although Steno was not the first to note the similarity. Fabio Colonna had postulated the same idea in 1616. Steno however developed these ideas and further noted through experimentation that sediment could form by the settling of particles suspended in water, once the turbidity of the water had been allowed to cease. Sand placed in a glass of moving water was sufficient to keep the particles in suspension, but when the movement of water ceased the sand particles settled out as sediment.

His work in Italy also led him to recognise that the sedimentary strata were laid down horizontally and sequentially. Later processes within the ground were responsible for folding and contorting the strata into the present landscape. Then erosion undermined the strata and carried material down to the flood plains. But he also held that the evidence he found in the rock layers was consistent with the biblical Flood account, and that properly understood the nature of the geological evidence and Scripture would be found to be in perfect harmony. Accordingly, fossils were found on the tops of mountains because the waters had once covered the earth, and deposited the dying and decaying animals within the sediments followed by fossilisation.

Steno later offered two possible explanations to account for the biblical Flood. Following Jean Buridan, a fourteenth century philosopher, he suggested that the earth was cavernous and occasionally these caverns might collapse altering the centre of gravity of the earth. Thus the mass of water would be redistributed across the globe covering dry land. Alternatively, he suggested that reservoirs of water within the earth might occasionally be extruded due to heat and pressure.[2] Descartes, more a theoretician than a practical scientist, preferred the action of heat for the formation of sedimentary rocks, believing that the earth was formed as a hot molten body and later cooled into layers from the outside towards the centre. Descartes was looking for rational regularities to explain the earth's shape without reference to Scripture, but Steno again showed that Descartes was wrong, this time for attributing the formation of the sedimentary layers to the action of heat. Steno wrote his initial thoughts in *Prodromus to a Dissertation on Solids Naturally Enclosed in Solids* or *de Solido* that was published in 1669.[3] He went on to explain the difference between natural crystals that could be accounted for by laws of nature, and fossils and seashells that could only be formed as once living organism.

As noted already, Steno's approach to the study of geology was somewhat different to Galileo's methodology, being more in tune with the Augustinian approach. Steno submitted to Catholic authority and was willing to allow Scripture to speak on matters of science as well as theology, and believed Scripture would not be contrary to nature when properly understood. Steno's scientific study of anatomy and the rock strata then, and his dissatisfaction with Descartes' ideas, led him to conclude that Scripture was correct when it spoke of a global Flood. Because Scripture had so much to say about this event as an accurate historical record, he

---

[2] Cutler, *The Sea Shell on the Mountaintop*, pp. 105-122
[3] Steno, N, (Transl. by) Winter, J.G., *The Prodromus of Nicholaus Steno's Dissertation Concerning a Solid Body Enclosed by Process of Nature Within a Solid*, (English Version), New York & London: Macmillan Co. Ltd, 1916

believed the study of the earth and fossils could not adequately be completed without reference to both the book of nature and Scripture. Steno's early Protestant upbringing was being brought to bear on his scientific studies because of his affinity for Scripture.

Steno's explanations for the geological features of Tuscany in *Solido* was not perfect, and his more developed writings were subsequently lost, so it is not possible to know how he developed his ideas in later life. The details of his findings then can only be considered in the broadest terms. However, he recognised important distinctions, and Steno's model of stratification has provided a useful starting point for the study of geology. What is of interest is how Steno developed his scheme according to the Genesis account of Creation and the Flood, and how Steno used a methodology that allowed Scripture to speak fruitfully as an accurate historical record. Steno believed that nature might be silent at times while Scripture speaks; at other times nature can fill in the gaps where Scripture is silent. But Steno, like Galileo, believed that Scripture and nature would not be in conflict if properly understood. There was apparently some difference in emphasis between Steno and Galileo. This may be considered a Stenonian approach to the harmonisation of Scripture and geology. This is also an approach used by subsequent Flood geologists during the nineteenth century, and in the present day.

Steno divided the stratified geological evidence into six time periods as identified in the rock layers of Tuscany, and then by extension he applied them universally across the globe. Two aspects were identified when water covered the land, two aspects when the land was level and dry, and two when the land was broken and folded. Steno set forth his view of the agreement of nature with Scripture by addressing chief difficulties in terms of these six aspects. He noted that in regard to the first aspect Scripture and nature are in agreement when the whole earth was covered with water, but noted that nature is silent on the mechanism and timeframe. However, Scripture is able to relate information where nature is lacking. Steno further comments about the second aspect, writing that

> ...Nature is likewise silent, Scripture speaks. As for the rest of Nature, asserting that such an aspect did at one time exist, is confirmed by Scripture, which teaches us that the waters welling from a single source over-flowed the whole earth.[4]

On the third aspect Steno noted that neither Scripture nor nature provide an account and both leave uncertain the exact timeframe of the

---

[4] Steno, *Prodromus*, 1916, p. 264

formation of mountains. Nature he commented only proves that there was great unevenness, while Scripture makes mention of mountains at the time of the Flood.

> But when those mountains, of which Scripture in this connection makes mention, were formed, whether they were identical with mountains of the present day, whether at the beginning of the deluge there was the same depth of valleys as there is today, or whether new breaks in the strata opened new chasms to lower the surface of the rising waters, neither Scripture nor Nature declares.[5]

However, while Steno recognises that neither nature nor Scripture can date the exact time and shape of formation of the mountains, he noted that the hills were formed by sea because of the marine evidence, and that nature does not oppose Scripture in this regard.

> The formation of hills from the deposit of the sea bears witness to the fact that the sea was higher than it is now, that too not only in Tuscany but in very many places distant enough from the sea, from which the waters flow toward the Mediterranean; nay, even in those places from which the waters flow down into the ocean. Nature does not oppose Scripture in determining how great that height of the sea was, seeing that…definite traces of the sea remain in places raised several hundreds of feet above the level of the sea.[6]

Steno's methodology in developing the science of geology then appears to be different from that of Bacon and Galileo. His initial findings in geology were based on observations first, and then he sought to develop a model of the Flood from a study of Scripture and nature. And while Galileo had removed some of the Aristotelian thinking that hindered the development of astronomy, Steno in effect undermined the Aristotelian thinking that was previously influential in geology. That is belief that there was a sort of inorganic lapidifying power at work in nature that formed fossils. For Steno, the world did not contain an impersonal soul or mystical powers, but did present a record of the activity of God if properly understood through careful study.

The latter part of the sixteenth century and early seventeenth century then led to significant developments in science in terms of astronomy, anatomy and geology and also in terms of scientific methodology. However, popular accounts of these events today seek to

---

[5] Steno, *Prodromus*, 1916, p. 265
[6] Steno, *Prodromus*, 1916, p. 265

show that there was a battle between science and Christian faith, but the truth is more complex and somewhat different. It was Christians who developed science believing that there was truth in both natural philosophy and Scripture. There was a battle, but it was between ideas left over from Greek thought and present during the Renaissance, and the developing science based upon reason and observation. The Christian researchers believed that truth could be found in Scripture and in scientific studies. However, a dualism remained between the spiritual and material in the methodology of science, and this has led to continued disagreement over how science should operate, even to the present time.

Athanasius Kircher and the Spontaneous Generation of Fossils

The Jesuit Athanasius Kircher had a considerable reputation across a number of fields and was widely regarded for his science, although he was still caught up in more mystical Renaissance thinking. However, generally speaking the natural philosophers moved to reject the Renaissance symbolic interpretation of nature in favour of a realistic, scientific one.[7] Although Steno's studies in geology were incomplete, his research led to the gradual rejection of Kircher's theory that held that a mysterious force was responsible for the formation and existence of fossils.

Kircher had lived an eventful life and escaped a number of close shaves during his travels, once being caught in a powerful earthquake while journeying in southern Italy. The towns around were left in ruin, but the billowing smoke and activity of Mount Vesuvius fired his imagination. In terms of his geological approach, he allowed for the possibility that fossil shells near the seashore were once living creatures. Also, he considered, some large fossil bones might have been the relics of giant humans, but he also held that others were formed and placed in mountain rocks by an unknown plastic force of nature. Candidate laws included spontaneous generation, astral emanations, and the mysterious force of magnetism. Kircher also had an interest in Egyptian hieroglyphics, which he believed had mysticism properties. There was also a commitment to geocentrism in his thinking in disregard of Galileo's work, but in support of Aristotelian science and Catholic dogma.[8] However, Steno believed that Kircher was wrong on the formation of fossils. Instead Steno posited that the carrying capacity of turbid water was sufficient to suspend and carry dying creatures, and the sediment in which they later settled out and were buried.

In other areas however Kircher did some very useful work, for instance producing a detailed scientific defence of Noah's Ark in *Arca Noë*

---

[7] McCalla, *The Creationist Debate*, 2006, pp. 40-41
[8] Cutler, *The Sea Shell on the Mountaintop*, pp. 69-72.

in 1675. In this he believed the size and shape of the vessel could contain the 300 then known species of birds and animals, including food provision and waste. Animals subsequently adapted to their present form through a sort of natural selection as the various kinds spread out after the Flood. For instance, he thought that deer and reindeer should be considered to be of the same kind. Other smaller creatures however, such as worms and insects, he believed arose through spontaneous generation, or 'panspermia,' an idea commonly held at the time because of the influence of Aristotle. Louis Pasteur later demonstrated the folly of this idea.[9]

Fossils and the Royal Society

In England Steno's scientific work was known to the Royal Society through regular letters that were read at the Society's meetings. The newly formed Royal Society was considered the leading scientific organisation in the world following the demise of the Italian *Cimento*. More than any other organisation it represented Bacon's concept of a scientific judiciary. Many members showed an interest in Steno's scientific work at first, but following translation of *de Solido* into English in 1671 by Henry Oldenburg, then secretary of the Royal Society, opinions became mixed. One member who agreed with Steno that fossils had an organic origin was the rather difficult character of Robert Hooke who was raised on the fossil-rich Isle of Wight. Hooke's investigation of fossil wood through a microscope for instance showed that it had the same structure as freshly cut wood and he also discovered the cellular nature of plants. Although Hooke believed the Biblical account of the Flood, he did not consider the Flood to be responsible for all fossils as he thought it too transitory. Hooke believed that the earth had also been shaken by powerful earthquakes, on account of the highly broken and folded layers, together with the evidence of fossils found embedded in rock layers that were elevated on the tops of hills. For Hooke, as with Steno, knowledge of the history of the earth could be recovered from the rock layers, although both believed that the book of nature and Scripture were in harmony. Hooke commented that

> There is no Coin can so well inform an Antiquary that there has been such or such a place subject to such a Prince, as these [fossils] will certify a Natural Antiquary, that such and such places have been under Water, that there have been such kin of Animals, that there have been such and such preceding Alterations and Changes

---

[9] Breidbach, O, & Ghiselin, M.T., 'Athanasius Kircher, (1602-1680) on Noah's Ark: Baroque 'Intelligent Design' Theory,' *Proceedings of the California Academy of Science*, Vol. 57, No.36, Dec. 2006, pp.991-1002

of the superficial Parts of the Earth: And methinks Providence does seem to have design'd these permanent shapes, as monuments and Records to instruct succeeding Ages of what past in preceding.[10]

In 1637 Francisco Stelluti, a colleague of Galileo, had previously argued along the lines of the Aristotelian position, that clay naturally turned to wood in the ground. However, against the majority view, the ever-unpopular Hooke was having none of it and instead insisted that wood turned to clay in the earth. As well as being in a minority, another problem for Hooke was that he preferred to describe his theories in public lectures, and was less keen to write his ideas in books and papers as others were doing. Neither was Hooke someone to seek out natural allies. When he read Oldenburg's review of *de Solido* he did not feel vindication, but instead accused Steno of plagiarism through alleged illicit contact with Oldenburg.

For many decades later other members of the Royal Society continued to argue that fossil shells had a non-organic mystical origin. Martin Lister, a member of the Royal Society, and one of the foremost authorities on seashells argued against the organic origin of fossil shells in an abrasive manner, commenting that the 'Cockle-like stones' had always been stones and were likely the result of 'shooting salts'. He dismissed in mocking tones those who disagreed with him as *dilettantes*, people having little knowledge and ability in science, but Lister's own experiments to confirm his thesis were to prove fruitless. However, fruitless experiments were not a problem for the irrepressible Lister who published his work in 1678 where he argued that the evidence for the non-organic origin of fossil shells was everywhere to be seen in nature. Seemingly Lister was preaching to the converted in the Royal Society where the majority supported his non-organic theory of fossil formation. A student of Kircher, another Jesuit priest and Aristotelian Renaissance thinker Filippo Bounanni, agreed with Lister and also argued that fossil shells were of inorganic origin, being the result of spontaneous generation. Bounanni could not imagine how the sea could have covered the mountaintops.[11]

As the passions of the Reformation were subsiding, a change in attitude towards nature was becoming more apparent with the violence of the Flood and judgment for sin giving way to more deistic attitudes. During this time consideration of nature's design showed the goodness and wisdom of God according to geometric laws and mechanical regularity. Accordingly, the biblical account of the Flood was being re-interpreted as a tranquil affair. In terms of science, Robert Boyle, who had a measure of

---

[10] Hooke, R., *Posthumous Works of Robert Hooke*, London: Published by Richard Waller, 1705, p. 321, in McCalla, *The Creationist Debate*. pp. 19-21
[11] McCalla, *The Creationist Debate*, pp. 134, 151-3

sympathy for Steno's views, argued in a very popular pamphlet that the bottom of the sea was tranquil and not given to violent revolutions and storms. This approach gained some sympathy amongst members of the Royal Society including Newton. Newton believed that ethereal matter was descending to the surface of the earth from the heavens and renewing the surface as food from the sun and planets. Majestic mountains he believed were evidence of Creation, not the Flood.[12] Kircher had also argued that mountains were created as things of beauty revealing a divine, everlasting plan of goodness.

Another naturalist who partly accepted Steno's organic origin of fossils, albeit equivocally, was John Ray. His main stumbling block was acceptance of the implications of Steno's work as a result of his interpretation of Scripture, and also from lack of a credible model to account for such a massive inundations of water required for a global event. For theological reasons Ray believed that no animal could become extinct, even though many fossil forms did not closely resemble living organisms in his day. An idea of Hooke's, that perhaps species could change over time, was equally abhorrent to many, although it was known that artificially selective breeding programmes could alter the outward appearance of animals such as dogs, cattle and guinea fowl.

In 1691 Ray published *Miscellaneous Discourse Concerning the Dissolution of the World*, in response to claims by Edward Lhwyd regarding giant boulders that littered the valleys of Snowdonia. Lhwyd had argued that it would have taken many thousands of years, much longer than the biblical timeframe allowed, for those boulders to fall in this way. At the time neither considered the after effects of the glacial formation of the 'U' shaped Llanberis and Ffancon valleys in Snowdonia, or periodic earthquakes.[13] Rate-uniformity does not take into account the action of un-recorded earthquakes in history, and cannot therefore be used with accuracy to account for the formation and rock distribution in the valleys. Especially when there is such uncertainty over what may have happened in the past, for instance with consideration of volcanic intrusions in this area, and highly folded sedimentary layers.

---

[12] McCalla, *The Creationist Debate*, p. 156

[13] Ironically, Michael Roberts, a modern geologist and theistic evolutionist, highlights this claim of Lhwyd for the long age of the valley, and although Roberts acknowledges the glacial formation of the valleys, he does not take this into account when he seeks to account for the numbers of fallen boulders according to rate-uniformity. Roberts, M. 'Intelligent Design; some Geological, Historical, and Theological Questions,' in Dembski, M and Ruse, M *Debating Design*, Cambridge University Press, Cambridge, 2004, p. 281

Ray, who was seen as the leading naturalist of his time, also wrote about the perfect design seen in nature and in the beauty of the mountains in *The Wisdom of God Manifested in the Works of Creation*. Mountains he thought were the place of origin of springs, and provided minerals and resources for mankind's use. Ray came to believe in the stability of geology where processes acted according to divinely given regularities, but Ray continued to be attracted to Steno's view that fossils were of biological origin, although the possibility of extinction was problematic to Ray. This led Ray to consider that vast periods of time might be necessary to account for organic fossil shells that were embedded in sea floor sediment, and subsequently raised high above sea level in mountainous terrain.

For Steno his later conversion to Catholicism led to a gradual abandonment of scientific investigations with interest instead lying in spiritual matters. He trained for the priesthood and soon rose to the position of Bishop, later being sent as a Bishop to Hanover at the request of the German Duke Johann Frederich. It just so happened that at this time Duke Frederich also employed Gottfried Wilhelm von Leibniz as a family biographer and court librarian, although Leibniz was more interested in pursuing interests in theology, mathematics and geology than writing the family history of his employer. Leibniz was a brilliant scientist and very astute; he often joked that he had applied for and been accepted into the secret society of alchemists from merely copying and using words and phrases from relevant books at random. His attempt was so successful that he was offered the position of secretary of the organisation. Leibniz was also interested in developing plans to re-unite the Protestant and Catholic churches once more, and discussed this with Steno.

Steno was therefore friendly with Leibniz, although by this time the Catholic Bishop was reluctant to even talk about science and geology with Leibniz, a fact that Leibniz found most disagreeable.[14] However, Leibniz found Steno's work on geology so convincing that he quickly abandoned Kircher's plastic theory that he previously believed could account for fossils. He saw in Steno's work a new science from which he could draw-out fresh conclusions relating to the origin of humanity, the Flood of Noah, and other truths from the Bible.[15] Later, after Steno's death, Leibniz tried to find Steno's more developed manuscripts, even organising a geological trip to Florence, but it could not be found. It was later learned that it been put into the hands of Holger Jacobaeus, nephew of Thomas Bartholin, for safe keeping after Steno became Bishop to the German people. But when Jacobaeus returned home at short notice his belongings, and the Steno

---

[14] Cutler, *The Sea Shell on the Mountaintop*, p. 162
[15] Paraphrased from a quote in Cutler *The Sea Shell on the Mountaintop*, p. 162

papers were supposed to follow, but much of it failed to arrive. Sadly, that was the last anyone saw of Steno's more develop geological manuscripts.

In Britain Thomas Burnet became the Anglican chaplain to King William III, and developed an interest in geology from travels in the rugged terrain of the Alps. The deeply scarred valleys, and great folds in the mountains are notable to anyone visiting the Alpine region with their broken rock layers lying at sharp angles to one another. In Burnet's four-volume book *Sacred Theory of the earth,* first published in 1680 and completed in 1690, he wondered how this confusion in the natural order could be explained in light of a perfect creation. Burnet envisioned an original perfect creation with the surface of the earth displaying the characteristics of an eggshell, and like the egg it could easily be cracked. The broken mountains were evidence he thought of this broken, cracked earth where waters were able to pour forth across the surface of the world as the Scriptures asserted due to human sinfulness. This was an interesting idea for 1680, especially in light of later evidence for plate tectonics, explained in greater depth as late as the 1960s.

Burnet claimed to be working according to Scripture, reason and ancient sacred authorities, but he was also committed to Descartes' view that natural laws operated as perfect regularities after the Creation week. Cambridge Platonist' thinking, where the plastic theory of fossil formation was still popular, also influenced his work. Thus, he had little interest in explaining the existence of fossils, nor for integrating Steno's work on sedimentation into his own. Instead, Burnet believed the twisted rock layers he observed in the Alps were originally laid down at Creation as perfectly uniform layers, but later wrenched apart by powerful forces during the Flood period. Although Burnet thought his work would silence the atheists it in fact drew criticism from many quarters. Ray commented that it was a 'chimera' or 'Romance;' Locke that it could not be reconciled with philosophy or Scripture. And Burnet did seem to have an imaginative way of stretching the meaning of Scripture to suit his cause.[16] Newton believed the present shape of the created order was perfect and created according to a divine mind. This led him to write to Burnet, arguing that most of the strata were laid down at Creation, and likened the surface of the world to the effect that beer would have if poured into a bowl of milk and left to dry. The curdled mixture would resemble the surface of the world Newton argued.[17] Burnet also discussed with Newton in private correspondence the possibility that the six days were long periods of time with an allegorical interpretation. Newton considered it a possibility that the earth's spin was slower in the past, thus making a day a longer period of time and that the

---

[16] Cutler, *The Sea Shell on the Mountaintop*, p. 170
[17] Cutler, *The Sea Shell on the Mountaintop*, p. 172

earliest days were of indeterminate length.[18] A follower of Newton, William Whiston, was more sympathetic to a changing world and sought to develop more scientific arguments, arguing for instance in 1696 in *New Theory of the earth*, that perhaps passing comets and other catastrophes had shaped the surface of the earth.

Burnet's ideas did not fit with the religious milieu of the seventeenth century. The Church hierarchy increasingly viewed nature and the earth as immutable and perfectly formed, reflecting the wisdom and beauty of God. Such careful design did not harmonise with the apparent chaotic, ruinous state that Burnet observed in the rock layers. But Burnet did gain support from less lofty positions and his book went through seven editions between 1680 and 1753, and in the following decades of the late seventeenth and early eighteenth century over five hundred books and papers on geology were published including detailed monographs. This period also saw the Glorious Revolution of 1688, a revolution that saw the Catholic monarch driven out of England. As a result there seemed to be a rash of works that sought to reconcile the Genesis account of the Flood with the newly emerging scientific research. There were a number of more speculative papers written as well that encouraged mockery across the whole field.

John Woodward and John Harris

Another influential researcher to argue in favour of the Flood along the lines of Steno was John Woodward. His *Essay Towards a Natural History of the earth* appeared in 1695. Woodward was a successful physician who had gained his medical knowledge through an apprenticeship, not through a university education. This gave him a strong sense of confidence in his own opinion without much consideration for the views of others. It also gave him an inflated ego and overbearing attitude, and his vanity and rudeness became well known across Europe. By the time of his mid-twenties he started to develop a keen interest in collecting fossils, and over the following few years had visited many quarries and fossil sites across England. His growing collection gathered fossils from North America, with interest coming from Newton, Locke and Lister. Lister went so far as to sponsor Woodward's membership of the Royal Society, although Lister later regretted his support finding him impudent and troublesome.[19] Unlike Lister, Woodward rejected the plastic theory for the origin of fossils and with his notable fossil collection he was able to

---

[18] Baxter, S. *Revolutions in the earth*, London: Weidenfeld & Nicholson, 2003, p. 54
[19] Cutler, *The Sea Shell on the Mountaintop*, p. 177

gain a respectful hearing from his peers. Woodward could show from evidence that they were of organic origin. As far as extinctions were concerned he simply asserted that there were none, and that all the organisms evident from fossils were still living somewhere in the deep ocean or distant lands.

As people began to read Woodward's work, it soon became apparent that there was a degree of overlap with Steno's *de Solido*. While gossip persisted amongst critics, John Arbuthnot went so far as to compare lengthy passages and pages of both, and wrote a pamphlet *An Examination of Dr Woodward's Account of the Deluge and, with a comparison between Steno's Philosophy and the Doctor's in the case of Marine Bodies Dug out of the Earth*. This showed the degree of plagiarism. Some passages were copied word for word with no acknowledgement to Steno. However, there were notable differences especially in light of Woodward's appeal to special miracles to explain the sediments. Steno on the other hand believed the Flood to have been a natural event.[20]

Woodward believed that at the time of the Flood, the Newtonian force of gravity had been miraculously suspended, and thus all matter would return to its original primitive confusion with the earth reduced to a mixture of soil and water. Following this destruction of the surface of the earth Woodward argued that God then reinstated the laws of gravity. The fossils and rocks were laid down according to their weight with the heaviest at the bottom and lightest at the top. Such a theory attracted much criticism, not least because the evidence did not seem to fit the pattern with many fossils in the lowest layers appearing lighter than those in the higher layers. Also the depth of water required to suspend such vast amounts of material was considered to be greater than the available water on the earth. Unlike Steno, Woodward did not seem to have thought much about the carrying capacity of flowing water. However, the main criticism laid against Woodward's ideas was his continued belief that the fossils were of inorganic origin arising according to the popular mystical theory of fossil formation. Many leading members of the Royal Society and other scientists including Lister and Lhwyd at the end of the seventeenth century could not come to terms with the idea that fossils had a biological origin.

One supporter of Woodward's geology however was John Harris who was determined in his support of Woodward's ideas. Harris was able to examine Woodward's growing fossil collection and claimed that Woodward even had a recent ammonite shell (probably a nautilus shell). This provided evidence that extinctions did not happen. Such was his enthusiasm that Harris asserted that only atheists and free thinkers would

---

[20] Cutler, *The Sea Shell on the Mountaintop*, p. 176

continue to argue for the plastic theory of the origin of fossils. By 1697, in defence of Woodward's work, Harris would write

> *And the better and more discerning part of Mankind agree that* those Propositions *are abundantly warranted by the* Observations, *and proved beyond any reasonable Contest, to those who can judge of a* Proof:...

> For as 'tis impossible to imagine how the Shells, Teeth, and Bones of Fishes, could ever get down to such *vast depths*, as we find they are every where at land...and entomb themselves in the Bodies of *Solid Stone*.[21]

However, Harris's crusade did not mean that all accepted Woodward's ideas, far from it. There was in fact a continued bitter disagreement within the Royal Society with many members continuing to support the non-organic origin of fossils, Martin Lister included. Cutler comments that even in 1717 there was continued disagreement over these matters, even spilling over into the theatres where a play characterising Woodward descended into a shouting match between factions in the audience.[22]

Woodward also struck up correspondence with Leibniz during the latter part of the seventeenth century. Leibniz had published various thoughts in his journal *Acta Eruditorum* commenting that the earth had undergone greater changes than many people thought. Leibniz was willing to give credit to Steno for the evidence for stratification and the origin of fossils. However, whereas Steno saw two major inundations of the sea (at Creation and the Deluge) together with other lesser catastrophes, Leibniz later saw the history of the earth in terms of progressive improvement with sudden watery deluges downplayed. In *Protoggea*, published in 1749 thirty-three years after his death, he had argued that a receding ocean had left its mark as evidenced by the sedimentary layers. Leibniz seemed fickle and changeable in his approach to science and Scripture. He also saw the new science of natural geography offering benefits for mineralogy and the mining and quarrying industries. In contrast, Woodward found Leibniz' Scriptural approach insufficient, and lacking the scientific knowledge of fossils and the practical insightfulness that Woodward could boast from his

---

[21] Harris, J., *Remarks on some late papers relating to the universal deluge, and to the natural history of the earth*. London: Printed for R. Wilkin, 1697, quote is from the Preface (A2), and p. 16 (emphasis in original). Paraphrased in Whitcomb and Morris, *The Genesis flood*, Baker Book House, 1961, p. 91
[22] Cutler, *The Sea Shell on the Mountaintop*, pp. 180-181

large collection. However, at the turn of the century and in the few decades following it was becoming increasingly apparent that acceptance of the inorganic origin of fossils was an unsustainable proposition.

Chapter Summary

During the seventeenth century there was a division between the supporters of Flood geology, such as Steno, where evidence was based on Scripture and the study of nature, and those arguing for an inorganic origin of fossils. As noted the main centres for belief in the non-biological origin of fossils were the Jesuit priests of Kircher and Bounanni, and also leading members of the Royal Society including Lister. Some such as Ray were not completely convinced by the organic origin of fossils because of the problem of extinctions, but remained undecided. The Flood supporters developed interesting and important theories, for instance that the strata were laid down as sediment, and that the fossil shells were the remains of once living sea creatures. Burnet's idea that the earth's topography was the result of catastrophic events cracking the surface, much like a cracked egg, has also been echoed in the twentieth century with the discovery of plate tectonics. But what the Flood supporters lacked was a sufficiently developed knowledge and understanding of processes occurring in the earth that could account for the amount of water. It was not known where the waters could have come from, nor where the waters went following the deluge, although Steno had proposed the idea that water had welled up from within the ground, and Woodward had argued for a temporary suspension of the laws of physics.

On the other hand, many scientists, including members of the Royal Society, atheists, freethinkers and a number of Jesuits, argued for the non-organic, mystical origin of fossils. This plastic theory, with its spontaneous generation of fossils, was though becoming increasingly untenable at the beginning of the eighteenth century, although some proponents could not accept the biblical Flood for a number of reasons. Firstly, the apparent lack of available water was a major problem, but increasingly they viewed natural changes on the earth in terms of Newtonian regularities and perfectly ordered mechanisms reflecting the goodness of God. A cataclysmic Flood, the result of divine judgment for human sin, did not fit with an increasingly deistic faith, and God's actions in the world were removed to a distant past. There was also continued affinity for the ancient Greek philosophers such as Plato, Aristotle and various hermetic works, and this only served to hinder the development of science, as the Galileo affair and belief in the plastic origin of fossils showed.

But what can be shown is that the Flood supporters, especially Steno, were careful observers of the strata as well as being committed to the biblical text. Supporters of Steno included Woodward and Harris. As Steno noted, sometimes nature could speak, at other times Scripture would speak truth about the geological history of the world. Combined, both could provide a more complete picture of past events. But it was seemingly a modified approach to science than that advocated by Galileo. On the other hand those who rejected the organic origin of fossils were themselves committed to ancient Greek texts with their pagan influence, or to an increasingly deistic faith, and this proved a hindrance to the development of geological science. On the other hand, the problem for the Flood supporters was explaining the origin of the volume of water necessary to cover the earth.

7.

Geology, Deep Time and the Enlightenment

At the start of the eighteenth century it was becoming increasingly apparent to rational thinkers that the fossils really were the remains of once living organisms, and the Aristotelian influenced plastic theory of fossil formation was becoming unsustainable as a credible theory. However, many could not accept that these fossils were deposited as a result of the Noahic Flood. Instead critics later found appeal in the cyclical nature of Greek and eastern mysticism. There was reluctance to give up the idea that nature had a hidden power over itself. But such was the effect that Steno, Woodward and Harris had on the emerging science of geology that the Anglican establishment was able to maintain the ascendancy of theistic arguments in geological science in Britain. In 1728, the Woodwardian professorship was founded at Cambridge in recognition of Woodward's contribution to geology. The Flood model of Woodward retained continued support amongst the Anglican establishment in subsequent decades, and the Tories also appealed to it in defense of monarchy. But during this period Christianity and Scripture were under continual attack. Anthony Collins wrote *Discourses of Free Thinking* in 1713 as an attack on the accuracy of the Old Testament, miracles and prophecy. Matthew Tindal also sought to present Christianity as a purely naturalistic religion in 1730 with his book *Christianity as Old as Creation* in which he viewed miracles as mere superstition. While these books were tolerated at the time, Thomas Woolston, with *Discourses on the Miracles of our Saviour* (1727-1729), touched a raw nerve by disputing the miracles of Jesus and the resurrection. As such he was fined for blasphemy, and languished in prison because of his inability to pay the fine. This pushed open criticism of Scripture and Flood geology underground in Britain. As a result deistic views of God, and criticism of the biblical text became hidden in works of natural theology.

Developments in France

However, in France there were further developments with researchers returning to studies of shell-enriched sediments, but also with increasing interest in pagan and eastern religion. There was also growing political turmoil with struggles between revolutionary forces and the monarchy, the latter was supported by a powerful religious-political establishment. The promotion of deep time and evolutionary ideas were part of this milieu, perhaps in part pursued to undermine the concept of the divine right of kings to rule.

René Réaumur studied the sedimentary layers around the town of Tours in France with its mass of broken shells. Shell fragments were so profuse that the farmers collected the shells, ground them down, and used them as soil fertiliser. In a paper to the Paris Academy of Sciences in 1720 *Remarks on some fossil shells of Touraine and their uses* Réaumur noted that the Falun layers consisted of around seven metres of evenly distributed sediment. He argued that instead of this being the result of a single watery event lasting no more than a year, the thickness of the sediment meant that it was evidence of an inlet of the sea that had slowly retreated to the coast. For Réaumur the shells were of biological origin, but he questioned the idea that a single event could have left such evidence. Neither did he have any desire to fit the evidence within Ussher's timeframe for the Flood, claiming that it would have required thirty to forty centuries for the sea to retreat to its present position. Unlike Woodward and Steno, Réaumur was not too concerned with Scripture even though he accepted the evidence that fossils were once living organisms. But instead of this leading to a modification of Steno and Woodward's theory, a total rejection of the Noahic epoch was growing in the French elite society, as was the desire for revolution.

Benoit de Maillet

Such was the view of Benoit de Maillet who wanted to rid geology of the Flood entirely believing the earth was of the order of two billion years old with a cyclical view of the universe.[1] De Maillet was well connected and as a public servant reported to the King's ministers. He presented his geological claims in terms of estimated calculations of the rate at which water was receding into the earth via vortices and extrapolated this back through time. A gradually receding ocean he also believed was evidenced from growing silt deposited in the Nile region. Thus he developed an early uniformitarian methodology, but there was also clearly influence in his calculations gained from his interest in Hinduism. De Maillet had travelled widely in Egypt and the Middle East as a French diplomat and studied the Egyptian pyramids and ancient pagan documents, thus developing an interest in the ancient eastern religions. The Hindu view of the world was expounded in the Puranic literature that asserted that the universe undergoes a continual cycle of creation, destruction and

---

[1] Repcheck, J., *The Man who Found Time: James Hutton*, Simon & Schuster, 2003, p. 99

recreation. The day of Brahma was said to last for 4.32 billion years, followed by a night of similar length.

The work of de Maillet was presented for publication in Paris in 1735, although draft copies had circulated in Paris as early as 1718 with gossip concerning its contents spreading through Parisian society.[2] It was finally published in the French language in Holland in 1748, three years after his death, as a dialogue between a Hindu sage and a Christian missionary.[3] This publication was under a thinly veiled pseudonym as *Telliamed, as conversations between an Indian Philosopher and a French Missionary on the Diminution of the Sea, and the Origin of Men and Animals.* Telliamed is of course a simple reversal of his name, and the character was given the task of presenting the authors more radical views. However, one may wonder in passing whether there is similarity between Telliamed, and the Babylonian half-fish half-man god of the chaotic sea *Tiamat*, especially bearing in mind de Maillet's obsession with mermaids and mermen he thought were still living in many parts of the world. As well as eastern religion, Descartes, and Fontenelle's work of 1686 *Conversations on the Plurality of Worlds*, had an influence on de Maillet.

Fontenelle's book was in the form of a science fiction novel that discussed alien visitors and vast periods of time. It would seem that the science of early modern France was tinged with vivid imaginations that stemmed partly from religious beliefs. In his own work, de Maillet had argued that a single flood could not explain the sediments observed during this time, and believed that the sediments and fossils were evidence of the receding ocean that had once covered the earth. Echoing the rhetoric of later Darwinists, his original work had appealed to chaotic forces (*le hasard*) as being responsible for creating and shaping the world with no need for God.[4] Stott points out that Charles Darwin even acknowledged influence from de Maillet's writing for a time in the 3rd edition of *Origins*, although it was later removed following criticism from Richard Owen.[5]

De Maillet believed that those who opposed the truth of his views were obstinate for reasons of Christian religious conviction, although in doing so he seemed to have ignored his own influence, and conflated his thoughts and beliefs with observation in the field.[6] De Maillet's edited work further held that the days of Genesis were long periods of time, and he developed an early theory of evolution where marine animals gradually

[2] Stott, R, *Darwin's Ghosts: In Search of the First Evolutionists*, London: Bloomsbury Publ. p. 116
[3] Stott, R, *Darwin's Ghosts,* p. 108, 130.
[4] Stott, R, *Darwin's Ghosts,* p. 118.
[5] Stott, R, *Darwin's Ghosts,* pp. 1-17, 113-114
[6] Cutler, *The Sea Shell on the Mountaintop,* p. 192

turned into terrestrial forms as the sea receded. Flying fish became birds and mermaids apparently turned into women. The Jesuit priest Abbé Jean Baptiste le Mascrier however had heavily edited the original manuscript before publication; the purpose was to make it more harmonious with Catholic doctrine before publication, although it still caused a great stir and controversy despite the editing. At the time there was a strong degree of fear amongst those seen challenging the French monarchy and religious authorities. However, Voltaire sarcastically criticised de Maillet for such absurd suggestions, although it would appear that his motivation was largely concern that de Maillet's receding ocean offered too much support to the Noahic Flood.[7] The work of de Maillet was however a bestseller and influential across Europe.

Another practical geologist of the time, Nicolas Boulanger an acquaintance of Rousseau, asserted that belief in the Biblical timeframe of six thousand years would pass as the fossil evidence proved the world was thousands of centuries old.[8] For Boulanger though he thought humanity was eternal and he developed an historical account of how people had observed terrifying and dramatic changes over the millennia. Although the biblical event had forced acceptance of the real origin of fossils, the timescale of a recent Flood was now under question. In 1746 Jean Etienne Guettard presented evidence of erosion and fossils from his own observations to the Paris Academy of Science, but did not develop it into a broader scheme of geology.

Comte de Buffon

Georges-Louis Leclerc, Comte de Buffon ignored the fossils altogether, still believing them to be of inorganic origin. He was elected to the Paris Academy of Sciences in 1734, and became the director to the prestigious scientific institution Le Jardin du Roi five years later. His chief work was published in 1749 as thirty-four volumes entitled *Histoire Naturelle,* in which he imagined that the earth and planets had been formed by a collision with a comet and the sun. Later he followed de Maillet in believing that a universal ocean had formed that then gradually receded to the present state. He also believed that the sea was migrating westwards very slowly, which over time had shaped the surface of the earth. However, Buffon's work was criticised by the faculty of Sorbonne in January 1751 as being against the creed of the Church. However, his cosmology involving collisions with comets to form the earth did not seem

---

[7] White, A. D. *A history of the warfare of science with theology,* Vol. 1. Reprinted New York: Dover Publications, 1960 (1896), pp. 56-62
[8] Cutler, *The Sea Shell on the Mountaintop,* p. 192

to be the problem. Instead, the idea that the mountains and valleys were formed by the receding ocean waters as a secondary cause, and not formed at Creation, was a problem for the faculty. Buffon was forced to recant to keep his position, but he continued to develop his ideas in private. In *Epochs of Nature* published in 1778 he argued that the earth was of the order of 75,000 years old.[9] This estimate was determined from experiments with cooling material such as iron and belief that the earth had gone through seven epochs of time. This was a small concession to Scripture in that the seven days were to be interpreted as seven long ages.[10] However, Buffon continued to speculate about the age of the earth in unpublished papers prior to his death with estimates rising to 10 million years.[11] Whereas Buffon's ideas were later developed along the lines of the 'day-age' theory, another Frenchman Isaac de La Payrère had proposed in the seventeenth century the possibility that once 'pre-Adamites' had live on the earth.[12]

Voltaire

Voltaire adopted the ideas of Kircher having himself studied *Mondus subterraneus* as a Jesuit student, and this led him to argue that fossils were of inorganic origin appearing through spontaneous generation, and that the mountains were permanent fixtures of nature. He found it astonishing that philosophers could not accept that ammonites were produced naturally in the earth, and rejected Réaumur's studies of the shell-rich layers of Tours arguing that the fossils could almost be seen to 'vegetate' if watched for long enough.[13] The other influence on Voltaire was Newton. Like other deists Voltaire believed the universe ran according to perfectly designed laws with machine-like regularity. Therefore catastrophic upheavals in the earth could not fit the patterns of nature. Andrew Dixon White later commented that Voltaire had a theological system to support, one that was in opposition to the biblical account of the Flood.[14] According to White, following news of fossil fish discoveries on hills in Europe in 1760, Voltaire went to great lengths to deny that they were real fossils and wrote multiple chapters in support of his deistic faith to argue that fossils were not from the Flood. Instead, he asserted that fossil

---

[9] Repcheck, *The Man who Found Time: James Hutton*, p. 101
[10] Mortenson, T. *The Great Turning Point*, pp. 26-27
[11] Baxter, S., *Revolutions in the Earth: James Hutton and the True Age of the World*, p. 55
[12] Cutler, *The Sea Shell on the Mountaintop*, p. 190
[13] Cutler, *The Sea Shell on the Mountaintop*, p. 195
[14] White, *A history of the warfare of science with theology*, p. 228

shells were left over from travellers, crusaders or pilgrims returning from the Holy Land. And fossil bones found between Etampés and Paris were perhaps bones that once belonged to the collection of an ancient philosopher, although it is hard to gauge how serious Voltaire was in making these claims. White commented that Voltaire used all his wit and wisdom towards this end as he was driven to oppose the geological investigations of his time out of the necessity of his deistic theology.[15]

The Development of Geology in Europe

In Europe during the eighteenth century a number of mining engineers began to study the rock strata and made early attempts at classification, and some of this was based upon Steno's earlier work. Johann Lehmann published his collection of observations in 1756 from investigations of mine workings in Prussia. Lehmann, a lecturer of mining operations in Berlin, categorised three types of mountains. Accordingly, the first type was primitive, generally poor in fossils, and severely folded and of high elevation, which he thought from before the deluge. Flatter, stratified mountains, with many more fossils and sequences of banded strata, were considered the result of the Flood, and he thought other mountains were possibly younger and formed later.[16] A contemporary of Lehmann from Prussia, George Fuchsel, considered that the different strata represented different epochs of time, having been evidently deposited horizontally. Lyell commented that Fuchsel wanted as far as possible to explain all the evidence for geological phenomena by reference to the agency of known causes.[17] His work contributed to later studies on rock layers, and although Lyell makes mention of the fact that he wanted to utilise presently known causes for their foundation, Fuchsel considered that the most recent layer were evidence of a great deluge.

One Italian inspector of mines during the eighteenth century was Giovanni Arduino. Working in the province of Tuscany for much of his career he later became Professor of Mineralogy in Venice. Arduino divided the rock layers into three divisions, and his theories and classifications had a large impact on subsequent geologists in Europe throughout later

---

[15] White, *A history of the warfare of science with theology*, p. 228

[16] Adams, F. D., *The Birth and Development of Geological Science,* Reprinted New York: Dover Publications, 1954 (1938), p. 374-478. See also: Zittel, K., *History of Geology and Palaeontology,* London: Walter Scott, 1901, p. 35

[17] In Ritland, R., 'Historical Development Of The Current Understanding Of The Geologic Column: PART I,' *Origins,* 8(2), Geoscience Research Institute, 1981, pp. 59-76. Quoting Lyell, C., *Principles of Geology*, Vols. I-IV, London, John Murray, 1834, p. 76

decades.[18] The three rock layers identified were the *Primary*, *Secondary* and *Tertiary* layers; the Tertiary is still used today for part of Cenozoic period.

Arduino observed that the Primary rocks were generally devoid of fossils, highly folded and therefore considered primitive. The Secondary layers were less folded and consisted of clay, marl, and limestone, and further identified because they contained many more fossils. Arduino considered the Tertiary layers to be much younger as they were rich in fossils, and he believed that many of these rocks were derived from material that had come from the Secondary series.[19] The proposal that the recycling of old rock layers to form new and higher layers in the geological column suggested that more than one epoch was necessary to account for the fossil and sedimentary evidence. In older rocks there was found deposition and burial of living organisms that had turned to stone. These old rocks were then broken up and eroded with material transported to a new location, and then the process of deposition and burial was repeated. Arduino however also considered a forth type of rock formed by volcanism. He thought that the earth had undergone repeated periods of upheaval and subsidence, with many revolutions and metamorphoses taking place in the earth.[20]

It was these theories and systems of classification of the rock layers developed by Lehmann, Fuchsel, and Arduino, which laid the groundwork for later eighteenth and nineteenth century geologists to follow. This schema was interestingly not that dissimilar to Steno's earlier division of the rock strata into three periods of deposition and formation, suggesting a degree of influence. The idea that the layers were laid down over varying time periods strongly influenced the leading geologists in later decades. However, it is notable that the biblical Flood did not completely disappear from sight amongst many geologists, and the fossil evidence, especially in the higher sedimentary layers of clay sand and gravel, was considered good evidence for the Noahic epoch. Scriptural geologists generally maintained that much of this evidence could be explained in terms of the inundation and subsequent recessional period of the Deluge.

John Whitehurst helped to develop geology along similar lines to his European counterparts in England in the late eighteenth century. He was an influential engineer, clockmaker and practical geologist who settled in the industrial town of Derbyshire. He became a member of the Lunar Society alongside Erasmus Darwin, James Hutton, Josiah Wedgewood and

---

[18] Conybeare, W. D. & Phillips, W., *Outlines of the Geology of England and Wales*, Reprinted London & New York: Arno Press, 1978 (1822) p. xiv

[19] Zittel, *History of Geology and Palaeontology*, p. 38

[20] Adams, *The Birth and Development of Geological Science*, p. 374

Matthew Boulton. In his studies he sought to develop geology as a practical science so that mining engineers could predict where coal measures and minerals were to be found in the earth. He published his writing in 1788, entitled *An Inquiry into the Original State and Formation of the earth.* As a Christian he was not antagonistic to the Bible and sought to connect Creation and Flood accounts with the geological evidence, arguing for instance that subterranean fires and volcanic action had shaped the earth, and that coal was of organic origin being the remains of forest vegetation. In the Appendix *General Observations on the Strata in Derbyshire* he describes in detail the structure of underground formations, interpreting the carboniferous coal measures as being deposited in successive layers.

Whitehurst commented in the *Preface* of his book that although many volumes had already been written on geological formations, the purpose of his work in studying 'Subterraneous Geography' was to make the science subservient to the benefit of human life. Whitehurst sought to base his studies on observations of nature and leave aside more speculative hypotheses about the origin of the world. He noted that although some hypotheses contained important truths, others were too speculative for the late eighteenth century. He asserted that he did not wish to point out the faults of other geological systems, but was willing to use those parts he considered useful for his own work. The purpose being to derive a system of geology from truly existing causes and laws of nature, which he believed the Creator had used to form the world.[21] While Whitehurst was interested in upholding the integrity of Scripture, such was the pressure from his progressive peers that he felt bound to interpret the evidence in terms of natural processes known from the present. However, Whitehurst's approach was essentially hypothetical-deductive, as opposed to the hypothetical-historical approach of the Flood geologists who considered the biblical account to be an actual event recorded by human observers.

During the latter years of the eighteenth century and early nineteenth century the study of geology was essentially divided into two camps. The first camp has become known as Neptunism, named after Neptune, god of the sea. This was based mainly on the work of Abraham Gottlob Werner, while the other was Vulcanism, or Plutonism, named after the mythological god of the underworld, and it had its foundation in Scotland under James Hutton.

Werner became a lecturer at the school of Mines in Frieburg in Saxony, Germany in 1775, and the small mining college quickly grew in size and influence. He was an intelligent, charming and charismatic teacher, although not without controversy. As a notable collector of minerals he

---

[21] Whitehurst. J, *An Inquiry into the Original State and Formation of the earth,* London, 1778; (Second edition, London, 1786); (Third edition, London, 1792)

gained support from his students who were gathered from across Europe, although others strongly disagreed. Many distinguished scientists came to hear him as well. Lyell considered him to be a great oracle of geology, but believed that his popular work held back the development of the subject for many years.[22] Lyell also considered him to be too bold in his assertions and excessively dogmatic as a sort of scientific pope. Werner was however very methodical and careful with the recording and description of his observations. He reworked the previous ideas of Lehmann and Fuchsel and developed a new system of classification for rock strata based on chemical and mineral composition. Werner's system of geology was then influenced in part by the ideas of Steno and Woodward, although without adherence to the biblical timeframe, and more in line with the long-age thinking of the French theoreticians such as de Maillet.

Werner's main idea was that all the rock strata were chemical or mechanical precipitates left over from the gradually receding ocean that once covered the earth. This deposition included igneous, metamorphic and sedimentary layers, which he thought could be described as five stratified layers, apparently from studying similarities in the rings of onions. His five divisions began with igneous at the bottom, which he thought a precipitate of the sea, then came the highly folded layers above. After this came the flat fossil-rich strata, followed by the sands and flinty gravels, followed finally by the more recent volcanic intrusions. His view was that the rocks were laid down as they are now found, either flat, tilted or folded.

Owing to his popularity and authority, many of Werner's students pursued the controversy throughout Europe even though his ideas were later considered to be inadequate during his lifetime. Although Werner acknowledged the action of recent volcanic activity, Hutton went on to argue that all rock layers were in fact volcanic or igneous in origin, even if subsequently uplifted and eroded. The weathered minerals were later washed down river, where settlement occurred in estuaries and on the seabed. This was in contradiction of Werner's assertions. The deist Werner taught and believed that the earth was of the order of one million years old, although many Christian writers thought his ideas sufficiently orthodox to allow for the global Flood. However, although Steno's system of geology was still visible in Werner's five-level model. Werner's geological scheme was also a move towards the ideas and concepts of Comte de Buffon and de Maillet where the sea was believed to have receded very slowly, and the age of the earth was extended well beyond the biblical events and timeframe.

---

[22] Lyell, *Principles of Geology*, p. 82

James Hutton and the Scottish Dimension

It is noteworthy at this point to mention common friendships and purpose between the French philosophers ahead of the French Revolution and those of the Edinburgh centred Scottish Enlightenment. Edinburgh was considered the Athens of the North because of the influence that Greek philosophy and mysticism had in the city. The deism of the French thinkers was similar to that of Scottish Enlightenment philosophers and American revolutionaries, with close personal contact between them. David Hume for instance personally entertained Benjamin Franklin in his house for three weeks in 1771, and Franklin had earlier been given a doctorate degree from St Andrew's University in 1759. It is thus possible that James Hutton knew of Franklin and the desire of the French Enlightenment philosophers to rid the world of any notion of the biblical deluge. Franklin had also been implicated as the instigator of the American Revolution following his humiliation at the hands of the London Privy Council in the early 1770s. Later he helped to ferment unrest in the colonies, helping to draft the Declaration of Independence in 1776, before becoming the American Ambassador to Paris.

As noted Voltaire was also deistic in his theology, and not so far in thought from the Scottish deists. He was invited to join the Parisian Masonic Lodge of the Nine Sisters shortly before his death, a lodge that also included as a member and Grand Master Benjamin Franklin. Franklin then is in fact a link between Hume and Voltaire, between the French and Scottish philosophers. Franklin was also an inspiration to the work of Malthus on population growth, where Malthus noted Franklin's observations that weak plants will be crowded out by the stronger.[23] Malthus' own father was also well acquainted with Rousseau following his exile from France in the years 1766-1770. Hume had studied in France at La Fleche between 1734-1737 where Descartes had studied. Baxter has further noted that Hume was an inspiration to the American Constitution through James Madison,[24] and James Hutton also studied in Paris and Leiden in Holland in the mid eighteenth century.

Later William Paley, writing in *The Principles of Moral and Political Philosophy,* tried to counter the growing libertarianism from revolutionary philosophies he observed in the late eighteenth century. He noted for instance that the political theory of Rousseau was evident

---

[23] Malthus, T.R., *Principles of Population*, 6th ed., Ward Lock & Co., 1826, p. 2. This is discussed in Bowden, M., *The Rise of the Evolutionary Fraud*, Bromley Kent: Sovereign Books, 1982

[24] Baxter, *Revolutions in the Earth: James Hutton and the True Age of the World*, p. 120

amongst agitators for unrest in Geneva, and Paley complained that the social politics of John Locke was prevalent in political disputes in Great Britain. Erasmus Darwin also came under suspicion for being a supporter of revolution in Great Britain, and Paley in *Natural Theology* in 1801 reiterated his ideas that the monarchical rule of law was divinely given according to the order of creation.

In terms of geology, James Hutton's views were radically different from Werner, and he published his first book on geology in 1795, entitled *The History of the earth*. Ten years earlier Hutton had written a paper for the Royal Society of Edinburgh outlining his controversial ideas.[25] Although he carried out careful fieldwork around the volcanic outcrops of Scotland, his ideas were also influenced by his religious beliefs, which were increasingly deistic. Hutton gained his deistic view of the world from Colin Maclaurin, his Edinburgh university lecturer in mathematics, and also from David Hume. Maclaurin had been a student of Isaac Newton, and Newton wrote a letter of commendation for Maclaurin praising him highly. Although Newton is often considered a theist, he was in fact interested in deism and Arian theology, and also sympathetic to alchemy and the study of secret messages hidden within the letters of Scripture.

Hutton however became accepted amongst the leading academics in Edinburgh and was a close friend and Sunday dinner guest of Adam Smith, and a twice-weekly guest of the high court judge James Burnet, also known as Lord Monboddo. Hutton was acquainted with Erasmus Darwin and occasionally used his house as a base for field trips.[26] Monboddo was strongly influenced by the ancient Greek philosophies of Aristotle and Plato and took an interest in the religions of the Egyptians and Hindus with its cyclical view of history and commitment to evolution. He had an affinity for pantheism and believed that man had evolved from the Apes. Some of this Platonism seems to have found its way into the scientific theories of Hutton where he considered it possible that the earth was an organised body. Along with Smith and Joseph Black, Hutton formed the Oyster Club as a place to discuss science while consuming fine food and wine, sometimes to excess.[27] Other Edinburgh Societies were noted for their riotous and licentious behaviour.

Another possible influence on Hutton's thinking was de Maillet's book that was published in the year that Hutton was in Paris studying anatomy and chemistry. De Maillet's popular work gave the developing

---

[25] Hutton, J. 'Theory of the Earth; or an Investigation of the Laws observable in the Composition, Dissolution, and Restoration of Land upon the Globe,' *Transactions of the Royal Society of Edinburgh,* Vol. I, Part II, 1788, pp.209-304, plates I and II
[26] Stott, *Darwin's Ghosts*, p. 170
[27] Baxter, *Revolutions in the Earth*, pp 120-121

science of geology a lengthy history with vast ages and a process of measuring time involving extrapolating from current rate uniformity. Although interestingly, and rather too conveniently, such measurements of time involving billions of years also fitted in with de Maillet's interest in eastern religion, especially Hinduism. Leibniz's posthumous book was also published in 1749 while Hutton was at the Leiben University, and Buffon's work published in the same year would have given Hutton much to think about regarding the influence of heat and water and the age of the earth. Hutton's geological thinking embraced the mechanistic view of nature that Newton had passed on to Maclaurin; for Hutton God was seen as the great designer of a mechanical world. He also saw in nature the possibility of a Platonic world soul with a constitution possessing reproductive powers. Hutton believed he could also see designing wisdom in nature

> We have now considered the globe of this earth as a machine, constructed upon chemical as well as mechanical principles, by which its different parts are all adapted, in form, in quality, and in quantity, to a certain end; an end attained with certainty or success; and an end from which we may perceive wisdom, in contemplating the means employed.[28]

Noting for instance the complexity of nature, such as the optics of the eye, Hutton retained the idea that an original contriver had knowledge of science. However, no place was left for a divine being to work miracles or bring about a catastrophic flood. Hutton's geological view was that most of the sedimentary rock layers were formed through a continuous regular process of erosion with material slowly moving from mountains to the plains and finally to the sea.

> The heights of our land are thus levelled with the shores; our fertile plains are formed from the ruins of the mountains; and those travelling materials are still pursued by the moving water, and propelled along the inclined surface of the earth. These moveable materials, delivered into the sea, cannot, for a long continuance, rest upon the shore; for, by the agitation of the winds, the tides and currents, every moveable thing is carried farther and farther along the shelving bottom of the sea, towards the unfathomable regions of the ocean.[29]

---

[28] Hutton, *Theory of the Earth*, pp.215-216
[29] Hutton, *Theory of the Earth,* p. 215

Hutton saw in this erosion the destruction of the earth as 'an end to this beautiful machine,' and he argued therefore that there must be another process that can renew the land. In arguing his case he appealed to an idea that could be seen as endorsing a pantheistic view of geology as a form of the Gaia hypothesis. Echoing the views of Hume's character Philo in *Dialogues Concerning Natural Religion*, Hutton wrote that the earth might be seen as an organised body with a constitution.

> But is this world to be considered thus merely as a machine, to last no longer than its parts retain their present position, their proper forms and qualities? Or may it not be also considered as an organized body? Such as has a constitution in which the necessary decay of the machine is naturally repaired, in the exertion of those productive powers by which it had been formed.

> This is the view in which we are now to examine the globe; to see if there be, in the constitution of this world, a reproductive operation, by which a ruined constitution may be again repaired, and a duration or stability thus procured to the machine, considered as a world sustaining plants and animals.[30]

Having established the basis of his argument, Hutton again echoed Hume, and provided two alternative scenarios. Firstly, that the world was made intentionally imperfect, or secondly that it is not the work of infinite power and wisdom. He comments that.

> If no such reproductive power, or reforming operation, after due enquiry, is to be found in the constitution of this world, we should have reason to conclude, that the system of this earth has either been intentionally made imperfect, or has not been the work of infinite power and wisdom.[31]

The land he later argued was initially raised up by the power of heat or volcanism over very long periods of time. He considered that basalt was formed from volcanoes and granite came from magma as a renewing process, but the whole process of erosion and uplift he thought would require immense amounts of time.[32] Hutton claimed that there is no written record that can account for the evidence of fossils within the rock layers,

---

[30] Hutton, *Theory of the Earth*, p. 216
[31] Hutton, *Theory of the Earth*, p. 216
[32] Hutton, *Theory of the Earth*, p. 215

although interestingly he did allow the Mosaic account to provide a history of mankind.

> Now, if we are to take the written history of man for the rule by which we should judge of the time when the species first began, that period would be but little removed from the present state of things. The Mosaic history places this beginning of man at no great distance; and there has not been found, in natural history, any document by which a high antiquity might be attributed to the human race. But this is not the case with regard to the inferior species of animals, particularly those which inhabit the ocean and its shores. We find in natural history monuments which prove that those animals had long existed; and we thus procure a measure for the computation of a period of time extremely remote, though far from being precisely ascertained.[33]

Hutton understood the presupposition in his thinking, that the only acceptable assumptions for geology are those that are known from the present day as equable and steady. As such his position forced him to conclude a vast length of time to account for the geological evidence.

> In examining things present, we have data from which to reason with regard to what has been; and, from what has actually been, we have data for concluding with regard to that which is to happen hereafter. Therefore, upon the supposition that the operations of nature are equable and steady, we find, in natural appearances, means for concluding a certain portion of time to have necessarily elapsed, in the production of those events of which we see the effects.

> It is thus that, in finding the relics of sea-animals of every kind in the solid body of our earth, a natural history of those animals is formed, which includes a certain portion of time; and for ascertaining this portion of time, we must again have recourse to the regular operations of the world. We shall thus arrive at facts which indicate a period to which no other species of chronology is able to remount.

> In what follows, therefore, we are to examine the construction of the present earth, in order to understand the natural operations of

---

[33] Hutton, *Theory of the Earth*, pp. 217-218

time past; to acquire principles, by which we may conclude with regard to the future course of things, or judge of those operations, by which a world, so wisely ordered, goes into decay; and to learn, by what means such a decayed world may be renovated, or the waste of habitable land upon the globe repaired.

This, therefore, is the object which we are to have in view during this physical investigation; this is the end to which are to be directed all the steps in our cosmological pursuit.[34]

Hutton here elucidated the concept that the present is the key to the past. While a practical geologist such as Whitehurst wanted to study present processes for the purpose of developing practical geology, Hutton was developing a general cosmological theory within the framework of his deistic and Platonic beliefs. However, Hutton recognised the marine origin of the limestone sediments, but according to his scheme, which rejected the Mosaic account *a priori*, there was no historical record that could tell us when the animals might have been buried in the rock layers.

We find the marks of marine animals in the most solid parts of the earth, consequently, those solid parts have been formed after the ocean was inhabited by those animals, which are proper to that fluid medium. If, therefore, we knew the natural history of those solid parts, and could trace the operations of the globe, by which they had been formed, we would have some means for computing the time through which those species of animals have continued to live. But how shall we describe a process which nobody has seen performed, and of which no written history gives any account? This is only to be investigated, first, in examining the nature of those solid bodies, the history of which we want to know; and, [secondly], In examining the natural operations of the globe, in order to see if there now actually exist such operations, as, from the nature of the solid bodies, appear to have been necessary to their formation.

...There are few beds of marble or limestone, in which may not be found some of those objects which indicate the marine origin of the mass. If, for example, in a mass of marble, taken from a quarry upon the top of the Alps or Andes..., there shall be found once cockle-shell, or piece of coral, it must be concluded, that this bed of

---

[34] Hutton, *Theory of the Earth*, pp. 217-218

stone had been originally formed at the bottom of the sea, as much as another bed which is evidently composed almost altogether of cockle-shells and coral. If one bed of limestone is thus found to have been of a marine origin, every concomitant bed of the same kind must be also concluded to have been formed in the same manner.[35]

Hutton's work was published about the same time as Werner's theories were being taught in Europe. According to Hutton's ideas volcanic activity brought rocks to the surface, these were then degraded and denuded by mechanical and chemical action of wind, rain and frost, and the suspension was carried down rivers and deposited as sedimentary layers in the oceans. These layers were then compressed into hard rocks by the action of heat, and later lifted up above the sea, and so the cycle he believed continued. For Werner, the primary rocks were also precipitated out of water while volcanic activity was caused by the burning of beds of coal deep underground. For Hutton the sedimentary rock layers were deposited as a result of erosion of volcanic rocks by the action of water, and then compressed back into basalt and volcanic intrusions by the action of heat and pressure. Hutton then gave geology the concept that the present is the key to the past, and claimed that scientifically speaking there could be found 'no vestiges of a beginning and no prospect of an end', although he did not deny the possibility that there might have been a beginning.

Two critics of Hutton were the Irish mineralogist Richard Kirwan and the Anglicised French geologist Jean-André Deluc, both naturalists. Both men took a fairly literal view of the Bible, but also objected to Hutton's theory because of poor scientific evidence, and as Kirwan believed, because of perceived atheism. Kirwan commented that Hutton's methodology rejected the educational value of Scripture towards geology and was therefore deeply flawed. Kirwan likened Hutton's approach to geology to the study of Roman history from buried artefacts with important texts excluded.

> ...*past geological facts* being of an *historic nature*, all attempts to deduce a complete knowledge of them *merely from their still subsisting consequences*, to the exclusion of unexceptionable *testimonies*, must be deemed absurd, as that of deducing the history of Ancient Rome solely from the *medals* or other *monuments of*

---

[35] Hutton, *Theory of the Earth*, pp. 219-220

antiquity it still exhibits, or scattered ruins of its empire, *to the exclusion of a Livy, a Sallust, or a Tactitus.*[36]

John Playfair later reworked and republished Hutton's ideas in his *Illustrations of the Huttonian Theory of the Earth* (1802), perhaps to make them less deistic and thus more acceptable to Christianity, but also out of belief that Hutton's work was difficult to read and was facing criticism. For instance in was criticised heavily by the *Encyclopaedia Britannica* that continued to argue for the biblical Flood in the late eighteenth century. Playfair rejected Kirwan's claim of atheism by arguing that Hutton's concepts were in harmony with natural theology. He pointed out further, that according to Hutton, the Bible only dealt with the history of man, which allowed for a much older history of the earth. Playfair claimed that Hutton viewed the Flood as a tranquil affair that left no mark on the earth, and, he argued, Hutton was only applying Newtonian mechanics to the study of geology as evidence of a well-designed scheme.

David Hume, Erasmus Darwin and a Power of Generation

Another significant influence on Hutton was the Scottish philosopher David Hume, and many of the concepts that Hutton applied to geological processes can be seen in comments put in the mouth of Hume's character Philo in *Dialogues Concerning Natural Religion.* Hume's conversational style of writing may be compared with that of de Maillet and Fontenelle with the main character Philo given the task of arguing from an Epicurean perspective. Order was claimed to assemble naturally from disorder due to natural forces acting on falling atoms over an eternity of time.

> And this...continued Philo...suggests a new hypothesis of cosmogony, that is not absolutely absurd and improbable...The continual motion of matter, therefore, in less than infinite transpositions, must produce this economy or order; and by its very nature, that order, when once established, supports itself, for many ages, if not to eternity.

> Suppose (for we shall endeavour to vary the expression), that matter were thrown into any position, by a blind, unguided force... If the actuating force cease after this operation, matter must remain

---

[36] Kirwan, R., *Geological Essays*, 1799, p. 5, cited by Penn, G., *Comparative Estimate of the Mineral and Mosaical Geology,* Vol. I, 1825, pp. 150-152, sourced from Mortenson, *The Great Turning Point*, 2004, pp. 57-64

for ever in disorder, and continue an immense chaos, without any proportion or activity.... If a glimpse or dawn of order appears for a moment, it is instantly hurried away, and confounded, by that never-ceasing force which actuates every part of matter.

Thus the universe goes on for many ages in a continued succession of chaos and disorder. But is it not possible that it may settle at last, so as not to lose its motion and active force (for that we have supposed inherent in it), yet so as to preserve an uniformity of appearance, amidst the continual motion and fluctuation of its parts? This we find to be the case with the universe at present. Every individual is perpetually changing, and every part of every individual; and yet the whole remains, in appearance, the same. May we not hope for such a position, or rather be assured of it, from the eternal revolutions of unguided matter; and may not this account for all the appearing wisdom and contrivance which is in the universe? Let us contemplate the subject a little, and we shall find, that this adjustment, if attained by matter of a seeming stability in the forms, with a real and perpetual revolution or motion of parts, affords a plausible, if not a true solution of the difficulty. [37]

While the character Philo, who is generally recognised as providing Hume's view in the *Dialogues*, is seen as an Epicurean, there are times when Philo proposes a more mythological and magical view of the world. For instance in Part VII of the *Dialogues* Philo asserts that the universe resembles more that of animal or vegetable than a human artefact, and that comets provided seeds for animal life. Therefore the universe by analogy should be considered more as a living organism.[38] Although Hume seemed to undermine the concept of miracles in his writing, and wanted to develop science along purely rational, empirical lines, he was closely associated with deists and Freemasonry. Such men included Erasmus Darwin and Benjamin Franklin with their interest in more esoteric beliefs. Erasmus Darwin for instance was initiated into St. David's Masonic Lodge No. 36 in Edinburgh in 1754, and was also a member of Canongate Kilwinning Masonic Lodge No. 2.[39] Erasmus Darwin's writing also appeals to Masonic

---

[37] See, Hume, D. *Dialogues Concerning Natural Religion*, (ed.) Smith, N.K, 2nd ed. Indianapolis, Bobbs-Merrill Educational Publishing, 1947, pp. 183-184. The text used here is from *The philosophical works of David Hume*, 1854.

[38] Hume, *Dialogues*, pp. 176-181

[39] Denslow, W.R., *10,000 Famous Freemasons*, 4 Vols., Missouri: Missouri Lodge of Research, 1957-61

belief, with for instance his assertion that *The Botanic Garden*, written in 1792, contained a structure that was based on the poetic Rosicrucian understanding of spirits in the elements. The *sylphs* represented air, *undines* water, *salamanders* fire and *gnomes* earth.[40] This also has echoes of Renaissance symbolic interpretation of nature. In *The Temple of Nature*, published in 1803, Erasmus Darwin also elaborated on the Eleusinian mysteries, in which he believed the philosophy of the works of nature were taught to the initiated by the Hierophant. These works he believed were first invented in Egypt, then transferred to Greece.[41] He believed that organic life had begun 'beneath the waves...' and that the 'first specks of animated earth' arose by 'spontaneous birth.'[42]

Such writing would suggest that he was proposing the worship of 'Mother Nature,' and indeed he wrote in these terms on occasions, for instance commenting in a letter about having explored the 'Bowels of old Mother Earth.'[43] However, officially he argued that his work was scientific and similar to that of Buffon, Réaumur, Ellis, Ingenhouz, Priestly and Girtanner.[44] Erasmus Darwin also seemed to suggest that Hume's own view was closer to that of pantheism, rather than Epicurean atheism, commenting that nature can be seen as possessing powers of generation.

> The late Mr. David Hume, in his posthumous works, [*Dialogues*] places the powers of generation much above those of our boasted reason; and adds, that reason can only make a machine, as a clock or a ship, but the power of generation makes the maker of the machine; and probably from having observed, that the greatest part of the earth has been formed out of organic recrements; as the immense beds of limestone, chalk, marble, from the shells of fish; and the extensive provinces of clay, sandstone, ironstone, coals, from decomposed vegetables; all which have been first produced by generation, or by the secretions or organic life; he concludes that the world itself might have been generated, rather than created; that is, it might have been gradually produced from very small beginnings, increasing by the activity of its inherent principles,

---

[40] The Eye (pseudonym), 'Erasmus Darwin Centre opens in Lichfield,' *Freemasonry Today*, Issue 9, Summer 1999. <www.freemasonrytoday.net/public/index-09.php>. Accessed March 2006

[41] Darwin, E., *The Temple of Nature, Or the Origin of Society,* London: St. Paul's Churchyard, 1803

[42] See Stott, *Darwin's ghosts*, p. 183, quoting E. Darwin *The Temple of Nature*, Canto I. 295-302.

[43] Stott, *Darwin's Ghosts*, p. 165-6. Letter to Richard Gifford 4th September 1768.

[44] Darwin, *The Temple of Nature*, pp. 3-4

rather than by a sudden evolution of the whole by Almighty fiat.— What a magnificent idea of the infinite power of THE GREAT ARCHITECT! THE CAUSE OF CAUSES! PARENT OF PARENTS! ENS ENTIUM![45]

The mention of a 'power of generation,' as making the maker of the machine, has similarity with Plato's 'source of generation' in the metaphor of the sun from the *Republic*. This power in nature was worked out through *Eros*, the Greek god of sexual love in Hesiod's writing (Hesiod was mentioned in Part VII of Hume's *Dialogues*). While in the *Timaeus* Plato describes the creator as a lesser deity, the Demiurge, a skilled craftsman or architect. Furthermore, Erasmus Darwin *imagined* millions of ages for evolution to arise. He wrote

> Would it be too bold to imagine, that in the great length of time, since the earth began to exist, perhaps millions of ages before the commencement of the history of mankind, would it be too bold to imagine, that all warm-blooded animals have arisen from one living filament, which THE GREAT FIRST CAUSE endued with animality...[46]

For Erasmus Darwin then everything was generated from small organic beginnings as 'living filaments' such as shellfish. Such was his enthusiasm for marine organisms that he adopted the scallop as his family crest with the motto *E. conchis omnia* (Everything from shells). This was even painted on his carriage door for a time, until fear of losing business and livelihood forced him to remove it.[47] The scallop was incidentally associated with the Roman god of love Venus, and Greek equivalent Aphrodite, who was said by Hesiod to have floated ashore on a scallop shell. Erasmus Darwin asserted that Hume was of the same mind regarding the origin of life. Although most Hume scholars would probably identify Hume as more Epicurean than esoteric, this view tends to overlook the dualistic nature of Greek thought where the powers of generation and the concept of nature can be acceptable to both the pantheistic and atheistic mindsets.

Hutton also worked out a scientific methodology presented in a book in 1794 that was very similar to that of Hume. According to Hume,

---

[45] Darwin, E. *Zoonomia; or the laws of organic life*, Vol. 1, New York, 2nd American Ed, from 3rd London Ed., corrected by the Author, Boston Thomas and Andrews. 1803: pp. 400-401
[46] Darwin, E. *Zoonomia*, Vol. 1, 1803: p. 397
[47] Stott, *Darwin's Ghost*, p. 168

experiences should be divided between *impressions* and *ideas*. But for Hutton experience begins with *sensation*; that is, how senses are stimulated by contact with objects and other agents. This is then followed by the organisation of various sensations, which form a *perception* and finally ideas are conceived and developed as *conception*. But Hutton realised that with the piecing together of historical changes within the earth direct observations were impossible and he was left only with *inferences*. One of those inferences was that processes taking place in the present time could describe historical changes, although in a moment of frankness he noted that the idea that natural geological changes were, and are, 'equable and steady' was a *supposition,* but he felt one that was reasonable.[48] This untested assumption later formed the foundation for Lyell's geological considerations.

Chapter Summary

As the developing science of geology entered the nineteenth century there were in effect two competing camps. It was these two camps of Neptunism and Plutonism that provided a basis on which nineteenth century geologists built. One, largely based on the localised geology of Scotland, promoted volcanism as a mechanism for the formation of new rocks, while the other, based on European geology, advocated a long process of sedimentation. Both ideas promoted belief that the earth was very ancient, at least a million years old in the case of Werner, and of indeterminate age according to Hutton's view. The two observed types of rock formation, formed by volcanic activity and a process of sedimentation, are clearly seen and can easily be accepted today, but the failures of each theory were evident as well. Werner's view was later seen to be in error, and Hutton's theory that constant very slow processes alone were responsible for rock formation, was also seen to be inadequate at explaining all the evidence. It was apparent at the time that catastrophic events helped to shape the world as well, and indeed volcanic rock formation is very much a catastrophic and unpredictable affair. Bearing this in mind we might ask whether it is possible from the evidence alone to infer any age for the earth from present observations. It may be noted that ideas of chaotic motion make the dating of catastrophic events very problematic from a scientific vantage point.

What is also clear is that a number of eighteenth century theoreticians from France and Scotland, such as Erasmus Darwin and de Maillet, were influenced by pantheistic or deistic beliefs from Greek

---

[48] Baxter, *Revolutions in the Earth*, pp. 130-131

thought or Hinduism. This religious influence seems to have had an impact upon their scientific work. As a result of this non-Christian religious influence, long ages, involving millions or billions of years of gradual change, were promoted prior to any proper scientific justification. There was little attempt to hide the fact that such beliefs were influencing scientific studies in these spheres in the eighteenth century, although subsequently in the nineteenth and twentieth centuries this strong influence has been largely downplayed by modern science. While a few secular scientists today recognise that the ancient pagan religions influenced the science of the early modern period it is clear that such esoteric beliefs were continuing to shape the natural sciences of the late eighteenth and early nineteenth centuries as well. The Aristotelian inorganic theory of fossil formation, or belief in spontaneous generation, can for instance be heard still echoing in Erasmus Darwin and Hume's consideration of a power of generation or vegetation at work in nature.

As will be discussed in the next chapter, Lyell later revived and reworked Hutton's uniformitarian theory, and Cuvier developed Werner and Arduino's aquatic theories of rock formation, carrying a belief in long ages into the new century, with support from Dean William Buckland. During the nineteenth century Auguste Comte, Charles Darwin and Thomas Huxley more carefully removed the influence of pagan religion from science with the move towards scientism, although there remained ambiguity in the concept of nature between atheism and pantheism. It may be argued as well that scientism, with its call for scientific authority, especially through the Royal Society, displays a political aspect of Plato's social order from the *Republic*. Scientism also enabled the development of the conflict hypothesis, by for instance Lyell, Babbage, Huxley and Dixon White amongst others. These scientists were arguing that appeals to Scripture were a severe impediment to the progress of science, despite the success of Flood geology in forcing acceptance of the organic origin of fossils. No longer was Scripture seen to provide educational or heuristic value to science, as it had done with Steno's work on early geology. The eighteenth century moves perhaps reflect the prejudice of the time, a *zeitgeist* if you like, although in part there may also have been an intelligently designed move in that direction through clever rhetoric and sophistry. Although the belief in mysterious inorganic powers at work in fossil formation was gradually lost, through the late eighteenth century the pagan belief in a power of generation at work in nature was being re-applied to living organisms as an early form of evolutionary change by Charles Darwin's grandfather. This is a significant observation.

8.

Developments in Geology in the Early Nineteenth Century

Georges Cuvier

One notable geologist who helped to develop the idea that successive floods was responsible for the geological record and fossil evidence was George Cuvier. He was Professor of Comparative Anatomy in the Museum of Natural History in Paris, and according to the English translation of his *'Essay on the Theory of the earth'* his work incorporated the Genesis Flood into the geological record as being the last great catastrophe. His main research was conducted through field studies of the geological layers within the Paris Basin, along the banks of the river Seine. This allowed him to look at how various fossils had become encased in the more recent sediment. These layers contained the remains of alternate marine and freshwater creatures, and he conjectured that this was caused by great inundations of the sea that had destroyed former worlds, followed by the flow of fresh water from rivers. Cuvier's theory, known as the 'Doctrine of Catastrophes,' held that successive and violent revolutions had occurred in the ground, and these had totally destroyed former worlds and life forms. Such destructive events were then followed by divine recreations. In this regard he disagreed vehemently with Lamark's evolutionary idea that there was one continuous genealogical family tree, and believed instead that life forms existed as separate discontinuous groups, much as Aristotle had proposed. Cuvier's influence was considerable, a result of a commanding personality that held centre court in the area of anatomy and the newly emerging science of geology. His ideas readily progressed and became accepted by the wider scientific community.

The Formation of the London Geological Society

In Britain William Smith was engaged in mapping the geological strata of England and Wales. He trained as a mine surveyor, but later became a canal engineer at the beginning of the nineteenth century. This was an exciting period for engineers with the industrial revolution in full swing, and highly regarded technicians were developing engines that ran on the apparent certain and universal clockwork principles that Newton and others had described. Initially Smith worked in the coal mines to the south of Bath and then advised the company on the building of canal systems in Somerset. Later he travelled more widely, working on various other road developments, and mining projects. As a result of his work and travels he became well acquainted with the underlying rock strata of Britain, and expounded through observations the idea that different fossilised rock strata could be identified across large geographical regions. What he discovered from the excavations was that rock strata containing similar fossils could be identified over wide areas. Smith assembled the data that he found, and worked it into the first geological map of Britain, which he published in the summer of 1815. He gave no real opinion as to how the rock layers might have been formed, but his work helped other scientists promote the belief that the strata were laid down over long periods of time. However, the Geological Society decided to publish its own map in 1821 and William Smith's work was left largely unsold. As a result, Smith faced poverty and ruin, although he was later admitted to the Geological Society through the oversight of Dean William Buckland, Adam Sedgwick and Roderick Murchison.

The Geological Society was formed in London in 1807 as a sort of after dinner discussion group, but later developed into a more serious scientific society. Buckland soon became a member, and was also appointed as a Reader in Minerology at Oxford in 1813. In the same year he began a geological tour of Britain with George Bellas Greenough, the first President of the Geological Society, in order to research the underlying geology. It is likely that Smith's own pioneering work was an inspiration to these men, and the Society's map appears to have relied heavily on Smith's drawings and graphs.

Buckland was both a notable scientist and theologian and attained the lofty position of Dean at Westminster by 1845. It is understandable therefore that he had significant influence in Britain at the time, both in the arena of geology and theology. This English Cleric had been rector for a time at the historic Anglo-Saxon church in Axminster, a small town in East Devon in which he was born. He was therefore well acquainted with Mary Anning the noted fossil hunter of nearby Lyme Regis, and had developed an early interest in geology.[1] Throughout his life he demonstrated the kindness and compassion of a genuine Christian gentleman, and was, for instance responsible for raising a pension for the Dorset fossil hunter when she became too ill to continue her enterprising work. He also showed gratitude to William Smith, recognising that he was indebted to both Smith and Anning for many fossil and geological discoveries.

Buckland was also noted for his eccentric lifestyle. He was reputed to have eaten his way through most members of the animal kingdom. Crocodile steak and battered mice were offered to guests for breakfast for instance, although they were spared roasted mole, which he found

---

[1] See Pulman, G.P.R., *The Book of the Axe*, 4th ed. Kingsmead Reprints: Bath, 1975, pp. 691-2.

revolting. The only thing that tasted worse, in his opinion, was the large bluebottle fly. Furthermore, such was his reputation for humour that sometimes his guests struggled to tell whether he was being serious when talking about scientific matters.

Buckland praised Cuvier for his *Essay* following its publication in England, and claimed that it vindicated the text of Scripture. Like Cuvier, he was a keen field geologist, and spent a number of years searching for evidence for the biblical Flood amongst the gravels at the top of the geological record. He further developed Cuvier's ideas of successive floods by claiming that the more recent strata, containing the bones of exotic mammals, were evidence of the Noahic event. After a lot of research amongst the uppermost rock layers, the gravels of southern England, and a particular Yorkshire cave at Kirkdale, Buckland published his own work in 1823, entitled *Reliquiae Diluvianae*, or *Relics of the Deluge*. The Yorkshire cave contained the bones of many mammals including tigers, bear, deer, rhinoceros, elephant and hyenas. How they became assembled into the same location was a mystery, and the entrance, at less than one metre wide, was too small for the elephant and hippo to have entered naturally. Buckland believed that hyena's were responsible for bringing the carcasses to this site, and even obtained his very own hyena to prove the point by feeding it with legs of beef, the purpose being to assess the teeth marks left with those found on the bones from the cave.

With the publication of the geological map of Britain in 1821, the gentlemen of the Geological Society had managed to identify many of the secondary rock layers. At the bottom was the Old Red Sandstone or Devonian strata, with subsequent layers labelled as Carboniferous Limestone, Coal measures, New Red Sandstone (Triassic), Jura Limestone (Jurassic), and the more recent Cretaceous layers of chalk and Greensand at the top. Much of the work in describing the geological column was complete by 1821, and it is possible to see the foundation of the new theory to rival the scriptural account. Of course this was not the direct intention of many of the researchers, and the whole enterprise was viewed as a perfectly valid process by a respected cleric who believed in the infallibility of Scripture. However, Buckland was also willing to read beyond the biblical account to a time before the world began, and to stretch the meaning of the Genesis account so as to fit the six days into long periods of time. Although Buckland was keen to promote Scripture, more literal clerics denounced his work because it suggested a fairly gentle event. They pointed out instead that Genesis called for a much more violent event. George Cumberland stated

The fountains of water contained in the great depths of the earth were broken up...Universal subsidence must have taken place. The operation must have been pretty rapid and immense layers of strata must have formed, filled up with debris of the broken surface...there was a sudden production of a thick sequence of rock![2]

Scriptural geologists of the early nineteenth century included Andrew Ure, Granville Penn, George Bugg, George Fairholme, John Murray, George Young and William Rhind.[3] However, their views were largely ignored, partly because sudden catastrophes and divine judgment did not fit with nineteenth century ideas on gradual progress in nature and society. But also because the acceptance of the deluge became caught up in a cultural battle involving a monarchical theory of government, deism and liberal politics. The nineteenth century Unitarian faith, being non-Trinitarian, held Jesus as merely a good moral teacher amongst other spiritual teachers. As such it tended to be humanistic and preferred liberal politics with focus upon worldly wisdom, instead of belief that knowledge, power and wisdom resided in divine authority. The Anglican clerics held that such authority descended through the monarchy with belief in divine justice. For deists however, nature was seen as good and the idea that it was somehow fallen into imperfection due to the sin of mankind was considered unpleasant. Newton had for instance seen in nature clockwork regularity that provided evidence for the benevolence of God. Therefore an over-emphasis on the principles of natural theology by the Unitarian deists led in part to rejection of evidence for the biblical Flood. The old earth creationists were also unkindly dismissive of scriptural geologists, perhaps believing they were an embarrassment to geological science. Professor Adam Sedgwick, Reader in geology at Cambridge, accused them of being full of bigotry and ignorance and hostile to the pursuit of natural science.[4]

The Flood was declining in popularity, and Cuvier was happy to fit Buckland's diluvia thoughts into his own work by 1826. Sedgwick too was in agreement with the concept of successive catastrophic floods with the last one being the fairly tranquil Noahic event. So the leading clerical geologist of Britain during the early nineteenth century held that the Noahic Flood was real, but they introduced previous, and numerous catastrophic floods to account for all the strata, with the earth considered much older than the literal biblical timeframe.

---

[2] In Cadbury, D., *The Dinosaur Hunters*, London: Fourth Estate, 2001, p. 77
[3] See for instance: Mortenson, T., *The Great Turning Point*, Master Books, 2004
[4] Sedgwick, A. *Discourses on Studies of the University*, 1834, pp. 148-151; also Mortenson, *The Great Turning Point*, p. 216

As noted, the London Geological Society was formed in 1807 as a geological dining club, a place for gentlemen to talk about geology as an after dinner event, initially meeting at the Freemason's Tavern in London. Of the thirteen original members, only George Greenough had any training in geology, paying for a visit to the Academy at Freiburg. Several other members were involved in medicine and chemistry. However, numbers quickly grew and the following year twenty-six Fellows of the Royal Society joined them. Ten years later the numbers were over 400, and by 1825 it was incorporated as the Geological Society with 637 members. This was a disparate group, and petty infighting, and rivalry within the Society made for often amusing, and enjoyable gossip amongst the chattering classes of the time. One Thomas Webster, curator of the Geological Society's museum, commented that his colleagues were 'a band of busy, jealous, active and revengeful witlings who have gained and kept their ascendancy partly from contempt, and partly from the indolence of others.'[5] The scientific credentials, and character of the early members were subject to scrutiny, but more serious researchers, such as Murchison and Sedgwick, rose through the ranks in latter years.

It may seem odd that the foundation of the Geological Society was so successful despite the infighting and lack of practical experience, as other specialist societies, in chemistry and botany for instance, were often much less successful. One early member, Sir Joseph Banks, in fact resigned from the London Geological Society because its unprecedented rapid growth was seen as a threat to his own Royal Philosophical Society. Many of the early members were not actively engaged in practical geology, but instead were members of Parliament, lawyers and doctors. It would seem that someone like the practical geologist William Smith was not invited to join at first because he was a practicing geologist and not fit for the purpose of the society, which was more metaphysical and political than scientific. Smith also did not fit the picture of an English gentleman. It is suggested by several commentators that there was in fact a political agenda, and Grinnell has identified Lyell and Babbage as being amongst key players in that agenda.[6] However, while there is evidence that Lyell and Babbage and others were working to an anti-monarchic political and deistic agenda within the group, it is less clear that the Geological Society was implicated directly in that agenda as a body. Buckland was for instance elected to

---

[5] In Winchester, S., *The Map that Changed the World*, Penguin Books, 2001, p. 283

[6] Grinnell, G. 'A Probe Into The Origin of the 1832 Gestalt Shift in Geology,' *Kronos: A Journal of Interdisciplinary Synthesis* Vol.1, 1 No. 4, Glassboro N.J.: Kronos Press, 1976, pp. 68-76.

serve as Secretary of the Geological Society and there was clearly an internal division that became more apparent in the 1830s.

In fact there were seemingly three camps represented in the Geological Society with strongly diverging views. First there were the old earth diluvialist such as Cuvier, Buckland, Sedgwick and Conybeare who were largely in control of the science of geology at the start of the nineteenth century. They believed in multiple catastrophes, the last of which was the Noahic Flood, and thus they tended to favour the Neptunism of Werner. Also represented were a number of young earth scriptural geologists such as Andrew Ure, although their influence was small, and finally there were the liberals who believed that all the geological evidence should be explained by gradual processes without recourse to the Mosaic account. They had an affinity for the volcanism of Hutton.

The Early Influence of Charles Lyell

It would be wrong to assume that Charles Lyell had no religious belief, he was apparently deistic and Unitarian in his belief, but at times appeared troubled by evolution because of its affront to Christian theology. However, Lyell's real motivation is not entirely clear, and his statements often appear contradictory, seemingly for instance directing people and influencing their thoughts behind the scenes through private correspondence in ways that are counter to his public statements. An example being his encouragement towards Darwin to write *Origins*, but not publicly supporting Darwin's ideas until the time was ripe in 1863.[7] There is evidence that Lyell was sympathetic to Lamarckian evolution in the 1820s, but was reticent in publicly supporting it because of lack of scientific credibility. Irvine comments that Lyell's second volume of *Principles* closely correlates with Darwin's later work in *Origins*, and that upon his return to England Darwin was surprised to find Lyell more sympathetic towards evolutionary ideas than he previously thought.[8] Lyell became closely involved in Darwin's work once the Beagle voyage was complete. For Lyell then there is apparent inconsistency between public statements and private actions throughout his life. Lyell is an enigmatic character, and there is strong evidence from his letters that he was working to a hidden agenda, apparently in support of deistic religion and the Whig political cause. When trying to piece together a person's life, it is usually possible to follow a consistent lead between actions and written statements.

---

[7] See for instance: Bowden, M., *The Rise of the Evolutionary Fraud*, Bromley Kent, Sovereign Books, 1982, pp. 92-99

[8] Irvine, W. *Apes, Angels and Victorians*, Weidenfeld and Nicholson, 1956, p. 43, 58

Sometimes people change their views, but, unlike Lyell, this is usually a consistent and well signposted change, as for example the conversion of Saul (later Paul) on the road to Damascus.

In 1825 the Whig George Poulett Scrope published *Considerations on Volcanoes,* in which he sought to attribute natural events that had previously been considered acts of God to volcanic activity. Thus he revived and reworked the volcanic geological theories of James Hutton. Volcanic activity seemed to provide such a useful natural mechanistic account of the geological evidence, perhaps over vast periods of time, that he argued that God was no longer necessary. However, at the time Scrope's book was seen as being too radical even for the London Geological Society and it was largely ignored. Scrope, who was the son of a wealthy London merchant, decided to pursue the liberal cause directly through a seat in Parliament; a seat he was able to buy for himself due to his family's wealth.

It seems that it was realised by Lyell that a less confrontational strategy was necessary to undermine the authority of the Anglican establishment, especially in light of the way that both Hutton and Scrope's work had been sidelined by the Anglican group. In the years 1825 to 1826 Fleming was also undermining belief in the Noahic Flood through promotion of the tranquil theory, although perhaps unwittingly he did not realise the effect such a theory would later have on Christian faith. Fleming argued that there had been a flood, but that there was no evidence for it in the rock layers that had been laid down over millennia, thus separating the science of geology from the Bible.

Like Scrope, Lyell was also seeking to re-work the theory of the deist Hutton in his three-volume work *The Principles of Geology.* Grinnell comments that Lyell took a much less confrontational approach in attempting to destroy the foundation of the monarchical theory of government than his predecessors.[9] In order to do this it was considered necessary to undermine the Moasic account of the Flood that spoke of divine judgment and sovereignty. Lyell went to great lengths to build his case and did not so much argue that the Mosaic account was wrong, but that it was essentially mythological and risked impeding the progressive development of geology as a science. He deliberately ignored the evidence of the rock strata, and Scrope wrote to Lyell in September 1832, following the publication of the second volume that

> It is a great treat to have taught our section-hunting quarry men, that two thick volumes may be written on geology without once

---

[9] Grinnell, 'A Probe Into The Origin of the 1832 Gestalt Shift in Geology', pp. 68-76

using the word 'stratum', If anyone had said so five years back, how he would have been scoffed at.[10]

It was indeed a feat to write two volumes on geology and ignore the evidence of rock strata that had been observed throughout Europe, and well catalogued and mapped in England and Wales fifteen years earlier by William Smith. Smith had published his maps between 1815 and 1817, as had the Geological Society with its own map. But Lyell was committed to Hutton's volcanism, and gradual processes over untold ages of time. Lyell also avoided mentioning catastrophism, and presented one hundred pages in his introduction to the first volume to argue that the deluge was really mythology, containing theological meaning only. Lyell ridiculed Flood geologists such as Scilla and European theologians as being too prejudiced and dogmatic, therefore guilty of holding back geology because of their insistence that the Mosaic account was responsible for marine fossils. Such a theoretical fallacy, he argued, had a more serious impact on geology than was present in any other science.[11] The fact that the Flood geologists, such as Steno and Scilla, had successfully argued for the organic origin of marine shells, against the pantheistic inspired plastic-force theory of fossil formation, was conveniently overlooked by Lyell's historical study.

Lyell elaborated on the history of volcanic uplift and the regular forces of erosion. Therefore his main line of argument was not really in science, but in providing a revisionist history of geological change believing such an approach was the only way to overcome Mosaic geology. Lyell painted the history of the geological record in terms of physical certainty thus seeking to render geological history as science giving it predictive power. Popper incidentally called this process 'historicism,' and although Marxism and Darwinism were Popper's main lines of criticism, Lyell's development of gradualistic geology was to become, and still is, an integral part of the Darwinian theory.[12] Lyell was not seeking to establish the idea of long ages in the geological record; that was already widely accepted by the Anglican geologists thanks to deists such as Werner and Hutton, and French and Scottish enlightenment thinkers. Instead, Lyell was directly seeking to remove the notion that there was *any* evidence for the Mosaic Flood in the geological record. He wanted to remove the last vestiges of geological evidence that held the most recent sediment as

---

[10] Letter from George Scrope to Charles Lyell 29th Sept. 1832, see Grinnell, 'A Probe Into The Origin of the 1832 Gestalt Shift in Geology' pp. 68-76

[11] Lyell, C. *Principles of Geology*, Vol.1, London; John Murray, 1867, pp.37-38

[12] Popper, K., *The Poverty of Historicism*, London: Routledge and Kegan Paul, 1957; See also: Fuller, S., *Science Vs Religion? Intelligent Design and the Problem of Evolution*, Polity Press, Cambridge, 2007, pp. 125-131

evidence of the Mosaic deluge from the minds of the old earth catastrophists such as Buckland and Conybeare. Such was his success in this that Fleming was upset, thinking that Lyell had stolen his tranquil theory and re-branded it under his own name. And worse, Lyell was using it for a godless purpose instead of Fleming's ecclesiastical goals.

It is said that the victor gets to rewrite history, and this often colours thoughts about the past. Lyell effectively re-wrote the history of geological change, and placed the deluge in a mythological setting. Therefore he thought that he could overcome the authority of the clergy that supported God and the monarchy, and allow the ascendancy of the liberal and secular political cause. In a private letter Lyell commented that 'I conceived the idea five or six years ago that if ever the mosaic geology could be set down without giving offence, it would be in an historic sketch,...' This long-term secretive plan was mention in part of a lengthy letter of 14th June 1830 explaining his plan of action

> I am sure you may get into Q.R. [Quarterly Review] what will free the science from Moses, for if treated seriously the party are quite prepared for it. A Bishop, Buckland ascertained (we suppose Sumner), gave Ure a dressing in the "British Critic and Theological Review." They see at last the mischief and scandal brought on them by Mosaic systems... Probably there was a beginning – it is a metaphysical question, worthy of a theologian – probably there will be an end. Species, as you say, have begun and ended – but the analogy is faint and distant. Perhaps it is an analogy, but all I say is, there are, as Hutton said, 'no signs of a beginning, no prospect of an end.' Herschel thought the nebulae became worlds. Davy said in his book, "It is always more probable that the new stars became visible and then invisible, and pre-existed, than they are created and extinguished." So I think. All I ask is, that at any given period of the past, don't stop inquiry when puzzled by refuge to a "beginning," which is all one with "another state of nature" as it appears to me. But there is no harm in your attacking me, provided you point out that it is proof I deny, not the probability of a beginning...I was afraid to point the moral, as much as you can do in the Q.R. about Moses. Perhaps I should have been tenderer about the Koran. Don't meddle much with that, if at all. If we don't irritate, which I fear that we may (though mere history) we shall carry all with us. If you don't triumph over them, but compliment the liberality and candour of the present age, the bishops and enlightened saints will join us in despising both the ancient and modern physico-theologians. It is just the time to strike, so rejoice

that, sinner as you are, the Q.R. is open to you. If I have said more than some will like, yet I give you my word that full half of my history and comments was cut out, and even many facts; because either I, or Stokes, or Broderip, [close friends of Lyell] felt that it was anticipating twenty or thirty years of the march of honest feeling to declare it undisguisedly. Nor did I dare come down to modern offenders. Try themselves will be ashamed of seeing how they will look by and by in the pages of history, if they ever get into it, which I doubt...P.S.... I conceived the idea five or six years ago that if ever the mosaic geology could be set down without giving offense, it would be in an historic sketch, and you must abstract mine, in order to have as little to say as possible yourself. Let them feel it, and point the moral.[13]

Lyell further reveals his determination to overthrow the Mosaic geology in a letter to Roderick Murchison in 1829 commenting that he wanted to throw out the whole of the biblical system, but do so in such a manner as not to cause offence.

I trust I shall make my sketch of the progress of geology popular. Old Fleming is frightened and thinks the age will not stand my anti-Mosaic conclusions and at least that the subject will for a time become unpopular and awkward for the clergy, but I am unafraid. I shall out with the whole but in as conciliatory a manner as possible.[14]

What can be identified then from this is a general plan of action by Lyell. It seems that he conceived the plan around 1825, probably following Scrope's failure with his book. The plan involved writing an historical sketch of inevitable progress while ignoring the scientific evidence for sediment and strata. This plan would be promoted over a period of twenty to thirty years before it could be declared openly, all the while pretending to be no enemy of the Anglican geologists who he was secretly seeking to undermine. These letters also show that for Lyell it was proof of a beginning that he rejected, although he thought it probable that there was one in much the same way as Hutton had argued. Lyell was so successful in promoting his cause that he was elected Secretary, and then President, of the Geological Society by the very people he privately planned to

---

[13] In Lyell, K, *Life, Letters and Journals of Sir. Charles Lyell*, London: John Murray, 1881, pp. 268-271
[14] Quoted in Mortenson, *British scriptural geologists in the first half of the nineteenth century- part 1: Historical setting*, pp. 225-226

overthrow. The artful manner of his election can be seen from the following letter he wrote to Gideon Mantell in 1831.

> My Dear Mr. Mantell – I have been within this last week talked of and invited to be professor of geology at King's College, [in London] an appointment in the hands of the bishop of London, Archbishop of Canterbury, Bishop of Landaff, and two strictly orthodox doctors, D'Oyler and Lonsdale. Llandaff only demurred, but as Conybeare sent him (volunteered) a declaration most warm and cordial in favour of me, as safe and orthodox, he must give in, or be in a minority of one. The prelates declared "that they considered some of my doctrines startling enough, but could not find that they were come by otherwise than in a straightforward, and (as I appeared to think) logically deducible from the facts, so that whether the facts were true or not, or my conclusions logical or otherwise, there was no reason to infer that I had made my theory from any hostile feeling towards revelation." Such were nearly their words, yet Featherstonhaugh tells Murchison in a letter, that in the United States he should hardly dare in a review to approve of my doctrines, such a storm would the orthodox raise against him![15]

Lyell reveals more of his thinking in 1832 at the King's lecture where he comments that Scripture itself will be all the better for being ignored as a source for geological inquiry as well as science.

> I have always been strongly impressed with the weight of an observation of an excellent writer and skilful geologist who said that 'for the sake of revelation as well as science – of truth in every form – the physical part of geological inquiry ought to be conducted as if Scripture were not in existence.[16]

Lyell did not consider the fact that it was the supporters of the scriptural text who forced rejection of the plastic theory of fossil formation from geological science. In this lecture Lyell carefully hid his liberal politics and deistic faith by declaring, with pretence, that being excluded from science would elevate Scripture. However, Scrope earlier revealed the liberal nature of the plan. In other words, the Mosaic Law was as much under attack as the deluge. Scrope praised Lyell's election commenting in a

---

[15] Sourced from Lyell, K. *Life, Letters and Journals of Sir. Charles Lyell*, 1881, pp. 316-317, also in Mortenson, *British scriptural geologists in the first half of the nineteenth century- Part 1: Historical Setting*, p. 226
[16] Kings college lecture 4th May 1832

letter dated 12 April 1831 that he found 'malicious satisfaction' in the election, as it would force the 'Bigwigs' to accept 'the new doctrine' out of 'compulsion' not from 'taste.'

> By espousing you, the conclave have decidedly and irrevocably attached themselves to the liberal side, and sanctioned in the most direct and open manner the principle things advocated. Had they on the contrary made their election of a Mosaic geologist like Buckland or Conybeare, the orthodox would have immediately taken their cue from them, and for a quarter of a century to come, it would have been heresy to deny the excavations of valleys by the deluge and atheism to talk of anything but chaos having lived before Adam. At the same time I have a malicious satisfaction in seeing the minority of Bigwigs swallow the new doctrine upon compulsion rather than from taste and shall enjoy their wry faces as they find themselves obliged to take it like physics to avoid the peril of worse evils. I feel some satisfaction in this.[17]

Although Lyell was elected Secretary of the Geological Society, the Whig take-over of the Geological Society did not happen all at once, but it was a gradual process with the slow assimilation and suppression of evidence that supported the catastrophic position. For instance Louis Agassiz's theory of ice ages of 1839 in support of catastrophism was not rejected but simply re-interpreted in terms of uniformitarianism.

Charles Babbage was also involved in the Whig plan. He was Lucasian Professor of Mathematics (1828-1839) with interests in geology, Unitarian theology and manufacturing, and a failed bid for a seat in Parliament. Babbage was involved in the campaign to undermine the ruling Anglican Tory elite, later writing an unofficial *Ninth Bridgewater Treatise.* His involvement in the clandestine liberal cause can be seen from a request for help from Lyell, but Babbage refused to write a review of Lyell's work on grounds that he was a known radical and his open involvement would therefore be damaging to the cause. He wrote

> I think any argument from such a reported radical as myself, would only injure the cause, and I therefore willingly leave it in better hands.[18]

---

[17] Letter from George Scrope to Charles Lyell 12th April 1831
[18] Letter from Charles Babbage to Charles Lyell 3rd May 1832, sourced from Grinnell, *A Probe Into The Origin of the 1832 Gestalt Shift in Geology*, pp. 68-76

Grinnell comments further on the nature of Babbage's radicalism and the nature of the cause, noting that there existed a combined effort by a group of people to discredit Paley and the Tory monarchists, this through an attack on its theological and geological foundations.[19] Babbage's *Ninth Bridgewater Treatise,* published in 1837, represented an attack on the theology of the Anglican establishment and specifically the work of Buckland and Kirby. Later, Babbage continued to develop the conflict theme between science and faith, writing in 1851 *Reflections on the Decline of Science in England,* where it was argued that wealthy amateur Anglicans and Tories had a stranglehold on science. Bearing in mind the way Buckland and Conybeare had helped Smith and Anning, this is not entirely fair. Thomas Huxley later took up a similar theme as Babbage; both had thought that scientists from a lower social position were being discriminated against, and such people were more deserving. But in effect they only managed to establish a new elitist hierarchy; one that was liberal minded and secular, and every bit as restrictive of discovery if claims contradicted their prior commitments to evolution and deep time.

Chapter Summary

This chapter has looked at the rise of the Geological Society and evidence that Charles Lyell was deceptively working to undermine the conservative Anglicans, particularly those sympathetic to the monarchical theory of government. Lyell's work was influenced by liberal-political considerations and a deistic faith. There was also a hidden attack upon the Mosaic Law, and a system of ethics based upon rights and duties. Instead, a number of liberals were in favour of a Humean or Epicurean ethical system based upon the sentiment of sympathy. Historically however such sentimentality in ethics leads to hedonism and the selfish abuse of other human beings, and ultimately it undermines objectivity in science. Although the allusion to paganism in the work of Hutton was not present in Lyell's work, the ambiguity in the concept of nature remained.

---

[19] Grinnell, *A Probe Into The Origin of the 1832 Gestalt Shift in Geology,* pp. 68-76

9.

Darwin, FitzRoy and the Beagle Voyage

Lyell's book was to have a significant effect on the scientific and political community over the following decades despite its bias. Darwin was able to read Lyell's work while aboard *HMS Beagle*, and later upon his return to England Lyell went out of his way to befriend the young Darwin. He seems to have found the attention emotionally rewarding and flattering. As a child Darwin had endured a difficult relationship with his overbearing and rather cold father, and his mother died when he was young. At first he seemed to want to live a carefree life of hunting, leisure and exploration. As a student of medicine and later divinity, Darwin had a first-class education, although not always applying himself fully. Darwin was though exposed to the latest thinking in geology and evolution while in Edinburgh. He became a member of the Plinian Society while in Edinburgh in the years 1825-27. This society was named after Pliny the Elder, and was formed to discuss natural history and undertake field trips and studies in geology without the authority of the professors. It was while in Edinburgh that Darwin became acquainted with the atheist Robert Grant. Grant was then the senior member of the Plinian Society and a lecturer in an Edinburgh School of Anatomy, having travelled on the continent and studied in Paris. Darwin later wrote that Grant expressed his strong appreciation of Lamarckian evolution while out walking.

> He one day, when we were walking together, burst forth in high admiration of Lamarck and his views on evolution. I listened in silent astonishment, and as far as I can judge, without any effect on my mind. I had previously read the *Zoonomia* of my grandfather, in which similar views are maintained, but without producing any effect on me. Nevertheless it is probable that the hearing rather early in life such views maintained and praised may have favoured my upholding them under a different form in my *Origin of Species*. At this time I admired greatly the *Zoonomia*; but on reading it a second time after an interval of ten or fifteen years, I was much disappointed, the proportion of speculation being so large to the facts given.[1]

---

[1] Barlow, N. (ed.), *The Autobiography of Charles Darwin,* London: Collins, 1958, p. 49, in Keynes, R., *Fossils, Finches and Fuegians: Charles Darwin's Adventures and Discoveries on the Beagle 1832-1836,* London; Harper Collins, 2002, p. 7-8

Darwin also attended lectures by Robert Jameson and was exposed to the geological disagreement between Hutton and Werner, later coming to favour the Huttonian approach. The *Beagle* voyage must have seemed like a great adventure, although it was somewhat tempered by the overbearing Captain Robert FitzRoy who suffered bouts of anger and depression, intersperse with periods of warm friendship. As well as fulfilling the role of Darwin's captain, FitzRoy is also known for his work in meteorology and hydrology becoming the first Chief Executive of the Meteorological Office in Britain. But it is his relationship with Darwin, and contribution to the origins debate, that is of most interest here.

Captain FitzRoy and the Beagle Voyage

FitzRoy was born on 5 July 1805, at Ampton Hall in Suffolk, and was a direct descendant of Charles II and Barbara Villiers, the Duchess of Cleveland; thus explaining his family name FitzRoy, meaning son of royalty. He was also a nephew of Lord Castlereagh. FitzRoy entered the Navy at a very young age, and trained at the Royal Naval College in Portsmouth; formerly known as the Royal Naval Academy founded in 1733. He was a very capable student and applied himself fully gaining the distinction of being the first student to win the gold medal from the college with a 100 percent pass rate. This demonstrated an extraordinary scholarly ability and commitment,[2] and he was later to gain command of a ship at the young age of 23. This was mainly due to his ability, although his Royal lineage must have played a part in his rapid promotion, as did unfortunate events involving the suicide of the previous captain.

He therefore gained command of *HMS Beagle* in 1828. Three years later in 1831 he again took command of the *Beagle* as captain, thus beginning his most famous second voyage that carried Charles Darwin on expedition to South America and the Galapagos Islands. This voyage was to circumnavigating the world. FitzRoy at first was not that committed to Scripture, seemingly indifferent to faith and confused about the literalness of sacred texts, although later he became a devout Christian, added with a fearless outspoken character.

The original plan of FitzRoy for the second voyage had been to arrange a trip at his own expense to carry three natives back to Tierra del Fuego, together with two missionaries.[2] The three native South Americans had been brought back to England in order to educate them in the ways of Christian civilisation and then return them home. The planning for FitzRoy's second voyage was later taken over by the Navy and organised

---

[2] Gribbin, J. & Gribbin, M., *FitzRoy*, Review, Headline Book Publishing, 2003, p. 23. Comments by Professor Inman

by Admiral Francis Beaufort. It finally left Plymouth on 27 December 1831 after a number of difficulties including absent and drunk sailors. It was to last much longer than had been expected. The task given by the Admiralty was to survey the coastal waters of South America, this for the purpose of maritime safety in the treacherous waters and passages in this area.

Following recommendation from John Henslow, Darwin joined the Beagle as the ship's naturalist and gentleman companion of the captain. The plan was that while FitzRoy was examining the coastal waters, Darwin was tasked with surveying the surrounding geology, flora, and fauna. At first Darwin's main interest was geology, having conducted studies in Wales with Adam Sedgwick, but his interest in biology grew.

Return Home and Lyell's Initial Contact with Darwin

Lyell was very keen for Darwin to return from his global voyage of discovery, and when he finally made landfall in England, Lyell wasted no time in getting to known the young Darwin who must have felt flattered by the attention of the President of the Geological Society. He wrote that 'Lyell entered in the *most* good natured manner, and almost without being asked, into my plans.'[3] Keynes comments that Darwin and Lyell became close colleagues sharing each other's innermost thoughts and secrets throughout the rest of their lives.[4] After five years with the conservative FitzRoy, Darwin was drawn back into a more liberal world that included his brother Erasmus Avery Darwin. This brother was to become a member of the esoteric Cambridge Apostles. Also in his acquaintance was an early Whig feminist Harriet Martineau, who later translated the works of Auguste Comte into English, as well as Charles Lyell. Another influence on Darwin was Thomas Carlyle, a one time strict Calvinist who lost his faith over the problem of reconciling predestination and free will. This question later appears in Darwin's thinking in correspondence with Asa Gray.

Darwin first started developing his secretive notebooks in 1836 and at first Lyell was the only one to know of Darwin's work. It is likely then that the hidden nature of these notebooks was due to Lyell's insistence. Darwin's first notebook was devoted to geology; the second entitled *Zoonomia* was a reference to his grandfather's work. As discussed earlier there was the open espousal of pantheism in *Zoonomia*, and Bergman has

---

[3] Keynes, *Fossils, Finches and Fuegians*, p. 379
[4] Keynes, *Fossils, Finches and Fuegians*, p. 379

claimed that Charles borrowed heavily from *Zoonomia*, to the point of it being almost a direct copy in places.[5]

Narrative of the Beagle

Both FitzRoy and Darwin later wrote up the explorations of the Beagle in a three-volume work as the *Narrative of the Surveying Voyages of His Majesty's Ships Adventure and Beagle*. FitzRoy used material from Parker King and Pringle Stokes to edit the first volume, and wrote the second volume himself from his own material, while Darwin wrote the third. FitzRoy's account reveals that he first seemed unsure and doubtful of the accuracy of the book of Genesis in light of his observations, but it is noteworthy that the picture that FitzRoy presents is different to that of the Darwin revisionists who suggest that the voyage was instrumental in shaping Darwin's uniformitarian views. FitzRoy commented in the *Narrative*

> I suffered much anxiety in former years from a disposition to doubt, if not disbelieve, the inspired History written by Moses. I knew so little of that record, or of the intimate manner in which the Old Testament is connected with the New, that I fancied some events there related might be mythological or fabulous, while I sincerely believed the truth of others; a wavering between opinions, which could only be productive of an unsettled, and therefore unhappy, state of mind.[6]

FitzRoy gradually began to see the catastrophic nature of the sedimentary layers. Perhaps his later marriage to a devout Christian woman had some influence on him, but from evidence in the *Narrative* it seems that the geological evidence observed first hand was most influential in changing his view of Scripture. During the voyage at one time he commanded a trip up the Rio Santa Cruz in whaleboats to survey the river's course. This took place in April and May 1834 with Darwin a passenger. FitzRoy later reported his findings to the Royal Geological Society writing them up in the *Narrative*.

> Is it not remarkable that water-worn shingle stones, and diluvial accumulations, compose the greater portion of these plains? On

---

[5] Keynes, *Fossils, Finches and Fuegians*, pp. 385-386; Bergman, J, Did Darwin Plagiarise his evolution theory? *Technical Journal*, Answers in Genesis, 16(3) 2002, pp. 58-63

[6] In Gribbin and Gribbin, *FitzRoy*, p. 79

how vast a scale, and of what duration must have been the action of those waters which smoothed the shingle stones now buried in the deserts of Patagonia.... Though the bed of the river is there so much below the level of stratum of lava, it still bears the appearance of having worn away its channel by the continual action of running water. The surface of the lava may be considered as the natural level of the country, since, when upon it, a plain, which seems to the eye horizontal, extends in every direction. How wonderful must that immense volcanic action have been which spread liquid lava over the surface of a vast tract of country.[7]

FitzRoy here expressed amazement at the scale of the action of water on the sediment and it would seem that he was beginning to accept that catastrophic processes were important in shaping the landform. Evidence of seashells embedded in mountain rocks was also an important piece of information for FitzRoy in shaping his view of the literal nature of the Genesis record.

It appeared to me a convincing proof of the universality of the deluge. I am not ignorant that some have attributed this to other causes; but an unanswerable confutation of their subterfuge is this, that the various sorts of shells which compose these strata both in the plains and mountains, are the very same with those found in the bay and neighbouring places . . . these to me seem to preclude all manner of doubt that they were originally produced in that sea, from whence they were carried by waters, and deposited in the places where they are now found.[8]

In the *Narratives*, written up after the voyage, FitzRoy highlighted the geological evidence that 'Mr Darwin' had found in the Andes region of South America. Perhaps he was concerned with what was being said about the evidence, but FitzRoy seemed to want to set the record straight with his work. Darwin was said to have found evidence of fossils buried rapidly in marine sediment, in layers subsequently found lying thousands of feet above sea level. According to FitzRoy this pointed to a global Flood, continental scale erosion and massive uplift. It is this sort of evidence that was partly responsible for forcing FitzRoy to accept the reality of the biblical Deluge as a global event, and the sort of evidence that the Darwinists wanted to suppress.

---

[7] In Gribbin and Gribbin, *FitzRoy*, p. 155-156
[8] In Gribbin and Gribbin, *FitzRoy*, p. 163

One remarkable place, easy of access, where any person can inspect these shelly remains, is Port San Julian. There, cliffs, from ten to a hundred feet high, are composed of nothing but such earth and fossils; and as those dug from the very tops of the cliffs are just as much compressed as those at any other part, it follows that they were acted upon by an immense weight not now existing. From this one simple fact may be deduced the conclusions—that Patagonia was once under the sea; that the sea grew deeper over the land in a tumultuous manner, rushing to and fro, tearing up and heaping together shells which once grew regularly or in beds: that the depth of water afterwards became so great as to squeeze or mass the earth and shells together by its enormous pressure; and that after being so forced down, the cohesion of the mass became sufficient to resist the separating power of other waves, during the subsidence of that ocean which had overwhelmed the land. If it be shewn that Patagonia was under a deep sea, not in consequence of the land having sunk, but because of the water having risen, it will follow as a necessary consequence that every other portion of the globe must have been flooded to a nearly equal height, at the same time; since the tendency to equilibrium in fluids would prevent any one part of an ocean from rising much above any other part, unless sustained at a greater elevation by external force; such as the attraction of the moon, or sun; or a strong wind; or momentum derived from their agency. Hence therefore, if Patagonia was covered to a great depth, all the world was covered to a great depth; and from those shells alone my own mind is convinced, (independent of the Scripture) that this earth has undergone an universal deluge.

Proceeding to the west coast of South America, we find that near Concepcion there are beds of marine shells at a great height above the level of the sea. These, say geologists, were once under the ocean, but, in consequence of the gradual upheaval of the land, are now far above it. They are closely compressed together, and some are broken, though of a very solid and durable nature; and being near the surface of the land are covered with only a thin stratum of earth. They are massed together in a manner totally different from any in which they could have grown, therefore the argument used in Patagonia is again applicable here. But in addition to this, there is another fact deserving attention: namely, that there are similar beds of similar shells, (identical with living species) about, or rather below the level of the present ocean, and at some distance from it.

In crossing the Cordillera of the Andes Mr. Darwin found petrified trees, embedded in sandstone, six or seven thousand feet above the level of the sea: and at twelve or thirteen thousand feet above the sea-level he found fossil sea-shells, limestone, sandstone, and a conglomerate in which were pebbles of the "rock with shells." Above the sandstone in which the petrified trees were found, is "a great bed, apparently about one thousand feet thick, of black augitic lava; and over this there are at least five grand alternations of such rocks, and aqueous sedimentary deposits, amounting in thickness to several thousand feet." These wonderful alternations of the consequences of fire and flood, are, to me, indubitable proofs of that tremendous catastrophe which alone could have caused them; - of that awful combination of water and volcanic agency which is shadowed forth to our minds by the expression "the fountains of the great deep were broken up, and the windows of heaven were opened.[9]

Chapter Summary

There were a number of influences on Charles Darwin, including a close affinity for his grandfather's work with its pantheistic sympathy, some of which later appeared in *Origins*, and meeting with notable atheists during his time in Edinburgh. Undoubtedly Cambridge Platonism would have had an influence as well. Lyell also paid close attention to Charles Darwin's work upon return from the global voyage, encouraging him to begin making notes in a number of secretive notebooks. What we also find in FitzRoy's *Narrative* of the Beagle voyage is evidence that goes against the popular account of Darwin's trip. These popular accounts suggest that Darwin was driven by evidence from the trip that supported his move towards gradualism in geology. Instead the fossil evidence found in Patagonia, as outlined by FitzRoy, questions the uniformitarianism of Lyell and Darwin.

---

[9] Sourced from: Fitzroy R, *Narrative of the Surveying Voyages of HMS Adventure and Beagle 1826-1836,* A Very Few Remarks with Reference to the Deluge. Chapter XXVIII. Vol II, London: Henry Coburn. 1839, pp. 663-668

10.

An age of Dinosaurs

Moves towards macro-evolution and gradualism in geology were gaining ground in Britain in the first part of the nineteenth century, although the highly respected Anglican scientists continued to hold it at bay for several decades. Early theories of evolution appeared to challenge the authority of the Bible, and with it the moral and social fabric of British society. Many establishment figures felt threatened that Britain might fall into the same revolutionary chaos that had gripped France, and indeed Britain was in a state of unrest. In 1832 the Reform Bill redistributed power, giving industrial towns seats in Parliament, and increasing the size of the electorate. This small reform helped to move people away from political revolution. Owen and Buckland, and many other privileged scientists, took up the challenge to counter the growing problem of revolutionary and evolutionary theories. At the time it was relatively easy to hold evolution at bay because the promoters had inadequate theories. The evidence for evolution from the fossils was controversial and widely challenged. On the surface Charles Lyell did not see progress in the fossil record and was seemingly a firm opponent of Lamarck's evolution.[1] However, it would seem that Lyell was waiting for Charles Darwin, or someone else, to mould a more complete naturalistic theory of gradual evolution in support of his gradual geology, and his undeclared affinity for evolution.

The Anglican establishment geologists of the time came under intense pressure as a result of the liberal's activities. William Buckland, William Conybeare a close friend of Buckland, and Adam Sedgwick moved towards acceptance of the gradualism of Lyell. Sedgwick was an acquaintance of Darwin who took the young Darwin on a field trip to North Wales before the Beagle voyage, and also gave encouragement to Charles Darwin after his return; although later he became concerned about the ethical consequences of Darwin's theory.[2] Sedgwick wrote to Darwin with concern that Darwin had tried to break the link between humanity and accountability towards God.

> You have ignored this link, and if I do not mistake your meaning, you have done your best in one or two pregnant cases to break it. Were it possible (which, thank God, it is not) to break it, humanity,

---

[1] Forster and Marston, *Reason, Science and Faith*, p. 331
[2] Forster and Marston, *Reason, Science and Faith*, p. 136

in my mind would suffer a damage that might brutalize it, and sink the human race into a lower grade of degradation than any into which it has fallen since written records tell us of its history.[3]

While Sedgwick was concerned about undermining personal duties towards God, it suited the libertines and Whigs. At best they were naïve over the fuller consequences of moving towards subjectivity in ethical matters, and at worst willingly complicit in terms of the development of social Darwinism. At face value the Whig's wanted to free the people from subjection to the Church and the monarchical state, but in reality it gave *carte blanche* to scientists and capitalists to profit from the growing industrial revolution. And with the development of social Darwinism there was little concern for the welfare of exploitable workers. The reality of the industrial revolution meant an increasing belief in technological and scientific progress.

It is against this background, of a belief in the inevitability of progress in the technological sciences, that progress was written into the historical sciences of evolution and geology. All that was needed was a way of linking human technological development with biological development of the species, and finding evidence of such progress in the rock layers. It was also necessary to develop a mechanism to explain how evolutionary change might come about. Appealing to a plastic force to account for fossils, or simply ignoring the fossil evidence, was no longer credible for those who opposed the Mosaic teaching.

Progress in the Geological Record and an Age of Dinosaurs

Various anatomists working with reptile fossils seemed to provide evidence that organic life might have progressed through the geological record. In 1829 Gideon Mantell, a leading fossil collector and anatomist, claimed that there had been an *Age of Reptiles*, which preceded the age of mammals, and these giant quadrupeds were the 'Lord's of Creation' from a time before man had been on the earth. Deborah Cadbury comments that this supported the views of the evolutionists that there was a progressive process of evolutionary change from simple forms to the more advanced.[4] But Mantell was not entirely happy with the conclusion of the early evolutionists, and both he and others such as William Buckland and

---

[3] In Darwin, C., (ed. Darwin, F.), *Life and Letters*, Vol. 2, London; John Murray, 1887, p. 249

[4] Mantell, G., 'The Geological Age of Reptiles,' *New Philosophical Journal*, Vol. XI, Edinburgh, Apr-Oct 1831, pp. 181-185, in: Cadbury, D*., The Dinosaur Hunters*, London: Fourth Estate, 2001, pp. 171-175

Richard Owen opposed gradual biological progression on religious and philosophical grounds. For Buckland and Owen, the evidence from observation was also a powerful reason to oppose the progressive evolutionary theories of those such as Jean-Baptiste Lamarck and Geoffrey Saint-Hilaire. Mantell too struggled to come to terms with this progressive interpretation of the evidence.

Saint-Hilaire, for his part, had claimed that the extinct fossil reptiles had progressed through a process of evolution into more recent mammals, such as the *Megatherium*. Saint-Hilaire wanted to find evidence for *homologies*, or equivalence in parts between different kinds of animals, so as to provide evidence for common ancestry. He suggested, for instance, that the hard upper shell of some insects resembled the vertebrae of reptiles, fish and mammals. Accordingly, in a paper presented in February 1830 he suggested that the hard shells of some cephalopods, such as the nautilus and fossil ammonites, were equivalent to the backbone of vertebrate fish. This was highly speculative, and Cuvier easily and comprehensively dismissed his ideas, calling him a *poet*.[5]

Cuvier also roundly condemned the evolutionary ideas of Saint-Hilaire as 'pantheism' in a public speech on 8 May 1832 at the College de France. Cuvier gave an impassioned speech, highlighting *divine intelligence* as being behind organic life, but this brave stand took its toll on him. While many in the audience were overcome with emotion, Cuvier suffered a stroke as a result, and died six days later. With the death of the great Cuvier, people looked around for others to wear his mantle. To the dismay of Owen, Professor Grant, by this time an anatomist at the University of London, was proposed as the next Cuvier by the distinguished paper the *Lancet*. Thomas Wakley, the editor of this respected paper, claimed that Grant was unrivalled in the British Empire, but Grant favoured the evolutionary ideas of Lamarck and Saint Hilaire.[6]

Many establishment clergymen viewed the University College of London as a godless place in comparison to Oxford University, and London's King's College. However, Grant's lectures, in which he introduced and promoted the progressive evolutionary ideas of the Parisian evolutionists, were packed. Grant believed that as the earth cooled, the changing climate caused animals to adapt and change to their surroundings, perhaps even an ape might change into a man. As noted, Charles Darwin also knew Grant from the time Darwin spent in Edinburgh in 1826-1827.

---

[5] Cadbury, D. *The Dinosaur Hunters*, pp. 183-184. Cadbury relies on a work by Schneer, C.J., *Towards a History of Geology*, Ch. 2 by Bourdier, F., Boston Mass. MIT Press, 1967; and Rudwick, M.J.S., *The Meaning of Fossils*, Chicago; Chicago University Press, 1972
[6] Cadbury, *The Dinosaur Hunters*, pp. 184-185

The Bridgewater Treatises and Fossil Evidence

The Reverend William Kirby, a respected clergyman and one of the writers of *The Bridgewater Treatises*, believed Mantell's assertions to be flawed. Kirby claimed that fossils of the *Megalosaurus* had been found with those of the mammalian opossum, and also stated that reptile fossils had been found in various and more recent strata.[7] William Buckland also opposed the progressionists. In his contribution to *The Bridgwater Treatises*, he tried to explain the order of burial evident in the geological record. This he thought provided evidence of a whole series of divine creations, each destroyed in successive catastrophes. After all, the *Megalosaurus* and *Iguanodon* displayed perfect form, which was evidence of a superb designer. Perfect design was even evident in the pre-Carboniferous layers. The fossilised animals could not have progressed through trial and error, but reflected the action of an '...Intelligent Creative Power,' especially regarding the vision of trilobites, Buckland thought. He noted that

> In the *Asaphus caudatus*...each eye contains at least 400 nearly spherical lenses, fixed in separate compartments on the surface of the cornea.[8]

This provided powerful evidence in defense of divine creation and intelligent design, and yet Buckland appears increasingly confused with his writings in *The Bridgewater Treatise*. Trying to tie together the meaning of Scripture with varying interpretations of the evidence found in the geological record could not have been good for his health. He slowly developed mental illness (later diagnosed as tuberculosis). Buckland was attempting to hold together competing paradigms, and also hold at bay the ever encroaching deistic and atheist theories of progressive evolution. This in turn threatened social stability through violent revolution, but the subtle influence of Lyell upon his thoughts was telling. As far as geology was concerned, on the one hand he had the successive catastrophes of Cuvier, but all the while he seems increasingly influenced by the uniformitarian

[7] Cadbury, *The Dinosaur Hunters*, pp. 172-175, reference to Spokes, S., *Gideon Algernon Mantell, LLD, FRCS, FRS, Surgeon and Geologist*, London: John Bale and Sons and Danielson, 1927, p. 44, and to Kirby, W., *The Bridgewater Treatise*, Treatise VII, Vol. I, London: William Pickering, 1835, pp. 36-42
[8] Buckland, W., *Geology and Mineralogy Considered With Reference to Natural Theology, The Bridgewater Treatises*, 4th edition, Vol. 1, London: Bell & Dalby, 1869, pp. 334-335

geology of Charles Lyell who was seemingly undermining Buckland behind the scenes. One statement in *The Bridgewater Treatises* highlights this conflict concerning the 'Mosaic Deluge,' rock strata and extinct species

> Some have attempted to ascribe the formation of all the stratified rocks to the effects of the Mosaic Deluge; an opinion which is irreconcilable with the enormous thickness and almost infinite subdivisions of these strata, and with the numerous and regular successions which they contain of the remains of animals and vegetables, differing more and more widely from existing species, as the strata in which we find them are placed at greater depths. The fact that a large proportion of these remains belong to extinct genera, and almost all of them to extinct species, that lived and multiplied and died on or near the spots where they are now found, shows that the strata in which they occur were deposited slowly and gradually, during long periods of time, and at widely distant intervals.[9]

The statement above concerning 'slowly and gradually' deposited layers appeals to the encroaching ideas of Lyell, while the periodicity of 'widely distant intervals' looks back to Cuvier. There is a conflict here that is still not satisfactorily resolved in the minds of modern scientists. Either the rock strata were laid down very slowly over long periods of time, or suddenly in catastrophic events, as the two ideas are not reconcilable. Catastrophes such as earthquakes or tsunamis, which might lead to fresh deposition of material in the ocean and coastal boundaries, do not happen slowly. The lack of erosion between conformable layers also speaks against long periods between depositions. The evidence of rapid burial and good preservation of organic material, that would quickly be scavenged, also points to sudden burial due to the deposition of suspended organic material, clay and sand from water. And the evidence for rapid burial and preservation of organic material was not unknown to those such as Buckland and Lyell. Henry De La Beche for instance remarked around the same time concerning the Jurassic layers around Lyme Regis that

> Some of the fossils are so beautifully preserved, their bones so well connected, with even the contents of their intestines between their ribs, and with traces of skin upon them, that many Icthyosauri and Plesiosauri must have been suddenly enveloped alive, or immediately after death, by the matter of rock enclosed them, so

---

[9] Buckland, W., *Geology and Mineralogy Considered With Reference to Natural Theology*, Vol. 1, 1869, p. 13

that neither their decomposition took place in the water nor the predaceous animals existing in the same seas had access to their bodies. Fish also are so frequently found entire that we would adopt the same conclusion respecting their remains. So that while we suppose the layers to have been gradually accumulated, minor accessions to the mass from time to time may have been more suddenly caused. The number of Ammonites found under conditions from which we may suspect that the animal was alive and retreated into its shell when overwhelmed with mud, is considerable.[10]

Buckland also reported the ability to extract the purple dye *sepia* from fossilised ink-bags found with belemnite fossils in the Lias near Lyme Regis (and near South Petherton and Yeovil in Somerset). The fossil resembles the living *Loligo vulgaris*, and requires both sudden death and rapid burial for such organic material to remain. Buckland reported to the Geological Society in February 1829, noting later that

> I might register the proofs of instantaneous death detected in these ink-bags, for they contain the fluid which the living Sepia emits in the moment of harm; and might detail further evidence of their immediate burial, in the retention of the forms of these distended membranes; since they would speedily have decayed, and have spilt their ink, had they been exposed but a few hours to decomposition in the water. The animals must therefore have died *suddenly*, and been *quickly* buried in the sediment that formed the strata in which their petrified ink and ink bags are thus preserved. The preservation also of so fragile a substance as the pen of a Loligo, retaining traces even of its minutest fibres of growth, is not much less remarkable than the fossil condition of the ink bags, and leads to similar conclusions.[11]

Furthermore both marine and terrestrial organisms, including the dinosaur *Scelidosaurus* and fossil wood are buried in the same layers as marine *Icthyosaurs*, belemnites and ammonites. This is observable for anyone who spends time fossil hunting around Lyme Regis and Charmouth (a recent exhibit in the Charmouth, Dorset visitor centre shows an insect

---

[10] From De La Beche, H, *Report on the Geology of Cornwall, Devon and West Somerset*, Longmans: London, 1839, pp.229-230, also in Pulman, G.P.R. *The Book of the Axe*, 4th ed. Kingsmead Reprints: Bath (1875) Reprinted 1975, p.11,
[11] Buckland, W., *The Bridgewater Treatises, Geology & Mineralogy*, Vol. 1, 1869, pp. 256-259

wing buried next to an ammonite). The problem for Buckland and friends is that they seem to have ignored the carrying capacity of flowing water, and the way in which conformable layers can grow during deposition out of sometimes fast moving fluid. This evidence was known to Steno from experimentation and shown more recently through practical research by Guy Berthault.[12] This is also evidenced by the growth of layers following the Mount St Helens eruption in 1982. However, the early nineteenth century geologists were becoming increasingly confused and ignored the implications of the evidence, perhaps even to the point of being willfully blind in some instances. Buckland saw the evidence of common design between the fossil and the living form, but interpreted it in terms of successive catastrophes and re-creations because of influence of Cuvier.

However, in order to hold together the competing and confused geological arguments with Scripture, Buckland had to stretch the meaning of Scripture sometimes beyond breaking point. Although in so doing he was trying to counteract the growing progressive evolutionists in favour of an intelligent Creator, his willingness to adjust Scripture to his own ideas damaged his own defence. With regard to the meaning of Scripture in the Creation account, he considered it possible that 'millions of millions of years may have occupied the indefinite interval, between the beginning in which God created heaven and earth, and the evening or commencement of the first day of the Mosaic narrative.'[13] And perhaps the meaning of the word creation ("*asah*" in Exodus 20:11) may imply something different too '...it by no means necessarily implies creation out of nothing, it may be here employed to express a new arrangement of materials that existed before,' he wrote.[14]

Buckland suggested to Roderick Murchison that he should look for evidence for earlier rock strata below the secondary layers among the hills of Wales. Murchison took up this challenge and ventured to the Welsh border, where he was able to examine rocks from the transition layers. These red sandstones lay beneath the Carboniferous coal layers, and contained invertebrate fossils of trilobites, crinoids, and echinoids.

---

[12] Berthault G. 1986, Sedimentology—experiments on lamination of sediments, C.R. Acad. Sc. Paris, 303 II, 17, 1569-1574; Berthault G. 1988, Sedimentation of heterogranular mixture—experimental lamination in still and running water, C.R. Acad. Sc. Paris, 306, II, 717-724; Julien P, Lany, Berthault G., 1993, Experiments on stratification of heterogeneous sand mixtures, Bulletin of the Geological Society, France, 164-5, 649-660.

[13] Buckland, W., *The Bridgewater Treatises, Geology & Mineralogy*, Vol. 1, 1869, p. 17. Reported in Cadbury, *The Dinosaur Hunters*, p. 192

[14] Buckland, W., *The Bridgewater Treatises, Geology & Mineralogy*, Vol. 1, 1869, p. 25-26. Reported in Cadbury, *The Dinosaur Hunters*, p. 192

Murchison, Secretary of the Geological Society, named the layer Silurian after an ancient tribe that had lived there several thousand years before. When Murchison presented his findings to the Geological Society, he hinted at a progression of life found in the layers. The Primary layers were free of fossils, while the highly folded and faulted transitional layers contained simple organisms. The Secondary system contained evidence of an age of reptiles, while the Tertiary contained remains of mammals.

At face value Lyell appeared unhappy with this stepped progressive idea because he considered the fossil evidence unreliable, and especially because of the lack of evidence from the fossilisation of vertebrate animals.[15] The main difference between the invertebrate life forms in the different strata is that the lower strata contain mainly benthic or bottom dwellers, while the higher are pelagic, or free-swimming organisms. However, benthic crinoids and echinoids, and other bottom dwelling invertebrates, find their way into the Secondary layers too, and are still extremely numerous today. In other words, to ascribe a progression of life from simple to complex through the geological record is to misunderstand the nature of the evidence because the same level of complexity is found in each, from the trilobite to the belemnite to the present day cuttlefish, as even Buckland noted. Many of these organisms are still present, and the evidence from a lack of vertebrate fossils in the transition layers is not evidence at all. The highly faulted and folded transition rocks are, in effect, the fossilisation of an ancient sea bed, as most of the organisms found therein are bottom dwelling aquatic life forms. Some mammalian bones have also been found, such as *Castorocauda lutrasimilis*, among the Jurassic reptiles. And evidence of rapid burial, including preservation of stomach contents and the skin of marine vertebrates, was also acknowledged by leading fossil hunters during the early part of the nineteenth century, something Lyell must have been aware of.

Richard Owen carefully studied Buckland's arguments in *The Bridgewater Treatises*, although his conclusions were somewhat different. Owen, it would seem, did not fully accept that each individual animal was specifically designed, but thought that nature may have adapted itself according to divinely given laws. However, he did accept, along with Buckland, that the ancient giant reptiles were in every part equal, if not superior to modern reptiles. Whatever the mechanism for organic change, the idea of progressive evolution was therefore wrong. He set about gathering anatomical evidence, and regularly disagreed with his competitor, the evolutionist Robert Grant. Owen married into the family of William Clift, and his career progressed rapidly to a Hunterian Professorship and

---

[15] Reported in Cadbury, *The Dinosaur Hunters*, p. 193

council member of the Zoological Society by April 1836. The shrewd Richard Owen was then in a position to undermine Grant from his respected position. He did this with characteristic efficiency and ruthlessness by simply vetoing Grant's nomination to the Zoological Society Council.[16] Grant's access to the best fossil specimens was denied and as a result his career declined, followed by the decline of his once packed university lectures. Lyell's gradualism and a more developed theory of evolution were effectively held at bay for another twenty or so years as Owen and Buckland continued to hold sway in scientific institutions. However, as their health and judgment declined, the liberal deists with their naturalistic, progressive ideas eventually managed to gain the upper hand.

Chapter Summary

This chapter has explored how belief in progress, with its influence in Unitarian deistic faith, led to increasing confusion in the minds of a number of Anglican geologists; this because the geological evidence does not in fact support the uniformitarian hypothesis. Evidence of catastrophic burial in the strata of Southwest England was known in the early nineteenth century, as described for instance by Buckland and De La Beche. There was also the promotion of gradualism by atheists, and other libertines who had an interest in political change. The leading Anglican clergy were concerned to maintain a belief in God's activity in creation, but their position was undermined. This was partly a result of their own compromise in interpreting Scripture, and out of a desire to fit in with the times. The progressives were also cleverer in rhetoric and sophistry than the more straight-laced Anglicans.

---

[16] Reported in Cadbury, *The Dinosaur Hunters*, pp. 197-198

11.

The Fall Out from *Origins*

Another major challenge to the Anglican ascendancy in science was the publication of Darwin's book *The Origin of Species* in 1859, followed by the famous Oxford debate between Thomas Huxley and Bishop Samuel Wilberforce the following year. Darwin had proposed a natural mechanism to account for all life forms. With evolutionary progression asserted his ideas provided a very useful tool for those who wanted to overthrow the God-fearing Anglican establishment.

Following the *Beagle* voyage, FitzRoy remained on reasonably friendly terms with Darwin and visited him at Down House in Kent on a number of occasions; that is until after the last meeting in the spring of 1857. It would seem that their friendship became strained following the publication of Darwin's book in November 1859.[1] FitzRoy in fact began to publicly criticise Darwin's work. He also criticised the work of other scholars who were questioning the age of the earth and the accuracy of the biblical chronology. In this regard he started an exchange in *The Times* in December 1859 vehemently disagreeing with the dating of stone tools found near the river Somme. These had been dated to 14,000 years before present. This exchange was carried out under a pseudonym of *Senex*, which comes from the Latin *nemo senex metuit louem*, meaning, 'An old man should be fearful of God.'[2]

Christian Challenges to Darwin - The Oxford Debate

A famous debate took place in an upper room of the Oxford Natural History Museum in June 1860, about six months after Darwin had published *Origins*. The main speakers were Bishop Samuel Wilberforce and Thomas Henry Huxley. The Bishop, the son of the anti-slave campaigner William Wilberforce, was not an accomplished scientist although not absolutely incompetent in matters of science either. For the debate he relied on briefings from Richard Owen, the founder of the Natural History Museum in London. The meeting was organised by the British Association and was well attended with estimates of around one thousand people in the long narrow upper room. Exactly what happened is subject to debate, but things seemed to turn unpleasant as tempers flared.

---

[1] Barlow, D., "The Devil Within: Evolution of a tragedy," *Weather*, Royal Met. Soc., Vol. 52 (11), 1997, pp. 337-341
[2] Gribbin and Gribbin, *FitzRoy*, p. 264

Huxley was apparently rude to the Bishop and used his ample rhetoric to win over the crowd, but FitzRoy also made a memorable contribution. At the end of the meeting FitzRoy is reported to have held a large Bible above his head and implored people to believe the word of God instead of Darwin. He also expressed regret for Darwin's fallacious work.

The official report in *The Athenaeum* merely records FitzRoy as saying that he 'regretted the publication of Mr. Darwin's book and denied Professor Huxley's statement that it was a logical arrangement of facts.'[3] FitzRoy seems to have been shouted down and Lady Brewster was so overcome by the heated atmosphere that she fainted and had to be carried out. Later FitzRoy commented in private correspondence that Darwin's work caused him the 'acutest pain.'[4] FitzRoy's contribution to the debate stuck in people's minds as Julius Carus in a private letter to Darwin some six years later comments that

> I shall never forget that meeting of the combined sections of the British Association when at Oxford in 1860, where Admiral FitzRoy expressed his sorrows for having given you the opportunities of collecting facts for such a shocking theory as yours.[5]

FitzRoy continued to discuss the difficulty that Darwin's work had caused him in private correspondence with Sir David Brewster, who was a co-founder of the British Association. Brewster was also a strong opponent of evolution and commented to FitzRoy that 'Darwin's book and the essays and reviews are most alarming proof of the infidelity and rashness of distinguished men.'[6] Some time later in another correspondence with Brewster, FitzRoy compared Darwin's theory of evolution to the 'beast rising up out of the sea [from Rev. 13] . . . opening his mouth in blasphemy against God.'[7]

---

[3] *The Athenaeum*, 14 July, 1860
[4] In Barlow, D., *The Devil Within*, pp. 337-341, sourced from: Origins of Meteorology: An analytical catalogue of the correspondence and papers of the first Government Meteorological Office, under Rear Admiral Robert FitzRoy, 1854-1865, and Thomas Henry Babington 1865-1866, of the successor Meteorological Office from 1867, primarily during its first two years under the Scientific Committee appointed by the Royal Society, and of registers of Instruments issued by successive Meteorological Offices from 1854 to c. 1915. Held at the Public Records Office, Kew, England
[5] Letter from Julius Carus to Charles Darwin, 15th Nov. 1866, In Gribbin and Gribbin, *FitzRoy*, p. 325
[6] In Barlow, *The Devil Within*, pp. 337-341
[7] In Barlow, *The Devil Within*, pp. 337-341

FitzRoy had in fact been a distinguished scientist in his own right, and was accepted as a Fellow of the Royal Society for his hydrographic and chronographic survey. He was also given the honour of becoming the first Chief Statist of the newly formed Meteorological Department of the British Board of Trade, an organisation that is now known as the UK Met Office. He is also recognised as a pioneer of the weather forecast and the development of storm warnings (the sea area Finnesterre was also later renamed after FitzRoy in his memory).

His concern for human welfare also developed out of a strong sense of Christian duty. This led to a desire to protect life, especially the lives of fellow sailors through the transmission of storm warnings, together with interest in lifeboat institutions. Much of his personal wealth was also expended in service to the country. However, in his later life he found himself under personal attack for speaking against Darwin's work, and his finances were getting low. The difficulties faced in developing weather forecasts and storm warnings were stressful enough, but he was undermined at the newly formed Meteorological Department by the actions of Francis Galton and others who were critical of his work. He was allegedly exceeding his brief in the dissemination of warnings and forecasts. FitzRoy sadly died under tragic circumstances probably brought about by stress from over-work. We can see however that there were notable dissenters from the work of Lyell and Darwin in the late nineteenth century.

Lyell and Darwin's Secretive Plan nears Completion

Lyell's 30-year subtle plan was virtually complete, and the liberals effectively established the science of geology and evolution on purely secular lines having driven appeals to religious texts to the margins. At first this was through charm, later that charm was to turn to fierce rhetoric and a reign of terror for those who objected. The liberal geologists effectively took the idea of a belief in progress that was growing in society and wrote it into the geological record so as to support the idea of progressive evolutionary change. Once a belief in progress was established in the geological record it was possible to then use it to reinforce belief in evolutionary progression in society as historically inevitable. Furthermore, as we have seen also from the eighteenth century, both deep time and evolutionary ideas grew out of an affinity for pagan and Hindu religious texts. However, although the evidence in the record showed some differences to the flora and fauna we know today, it was not progressive. Many of the animals from the deepest layers showed similar complexity to those now alive, as evidenced for instance by the complexity of the eyes of trilobites. Some of the dinosaurs were greater in size than present reptiles, and the record shows that animals appear suddenly, and then remain the same.

The claim that the fossil record is progressive allowed Darwin, Spencer, Huxley and others to build a universal naturalistic theory, which included mankind in the great chain of inevitable and gradual progressive improvement; this without the need for the sustaining grace of a personal Creator. By 1864, scientists were beginning to boast that science was at last free of religious dogma, by which they meant Christianity.[8] Thomas Huxley and colleagues in the newly formed secretive X Club wanted to pursue scientific knowledge without the restraint of religious thought. They were leading figures within the Royal Society, and the British Association for the Advancement of Science. The nine members of the X Club met before meetings of the Royal Society to control the agenda. Such scientists considered themselves the new priests to be looked to for life's answers, the philosopher kings of Plato's city-state Polis as outlined in the *Republic*.

Huxley portrayed theologians as being in conflict with science, the enemy of knowledge. He believed that atheistic and agnostic thinkers should develop politics and society based on the new science of geology and evolution. Science was therefore considered able to establish a united and universal cosmic principle based upon evidence and reason alone, although ultimately this is really *scientism* and is incoherent.

Charles Darwin also alluded to Plato in the final paragraph of *Origins* (6th Edition) adding mention of the 'Creator', although in doing so he was arguably comparing his belief in evolution to Newton's scientific work on gravity. Darwin wrote

> There is grandeur in this view of life, with its several powers, having been originally breathed by the Creator into a few forms or into one; and that, whilst this planet has gone circling on according to the fixed law of gravity, from so simple a beginning endless forms most beautiful and most wonderful have been, and are being evolved.[9]

Exactly what Darwin meant by acknowledging the Creator is open to question, but the reference to 'forms most beautiful' is an allusion to the

---

[8] Jensen, J.V., '*The X-Club*,' *British Journal for the History of Science*, 1970, pp. 59, 63-72, 179. In: Forster and Marston, *Reason, Science and Faith*, p. 309

[9] Darwin, C. R., *On the Origin of Species by Means of Natural Selection, or the Preservation of Favoured Races in the Struggle for Life*. 2nd edition, London: John Murray, 1860, p. 490

works of Plato.[10] This passage may also be compared with Erasmus Darwin's idea in *Zoonomia* of a 'power of generation' at work in nature. Darwin's 'Creator' then is perhaps more closely linked to the Demiurge of Plato. But we see the ambiguity at work in Darwin's thinking here, where on the one hand he seems to be arguing for purely natural causes, but then nature may be an impersonal undefined force with god-like power. This also perhaps reflects ambiguity in the concept of nature, between atheism and pantheism, as previously discussed. In the *Descent of Man*, published in 1871, Darwin attributed even man's morality and character to naturalistic processes, commenting that the first foundations of the ethical sense lies in social instincts that were primarily gained through natural selection.[11]

However, at times Darwin appeared sick with worry over the implications of his writing; too ill to attend the 1860 Oxford debate and at other times expressing 'horrid doubts' and 'jumbled' thoughts in private letters regarding his theory.[12] Perhaps his conscience troubled him for attacking Christianity. He also revealed that he was aware of the secretive plan of Lyell to undermine belief in the Deluge as an attack on Christian faith in a couple of letters in his later life, but linking it back to Voltaire. Darwin wrote

> Lyell is most firmly convinced that he has shaken the faith in the Deluge far more efficiently by never having said a word against the Bible, than if had acted otherwise...I have lately read Morley's Life of Voltaire and he insists strongly that direct attacks on Christianity...produce little permanent effect: real good seems only to follow the slow and silent attacks.[13]

> ...yet it appears to me...that direct arguments against Christianity and theism produce hardly any effects on the public, and freedom of thought is best provided by the [gradual] illumination of...minds, which follow from the advance of science.[14]

---

[10] Stephen Snobelen discussed the Platonic phrase 'forms most beautiful' in the work of Newton and others, in an invited paper delivered at the Ian Ramsey (Oxford) conference, 'God, Nature and Design,' 10-13 July 2008. Paper was entitled, 'This most beautiful system': Isaac Newton and the design argument.'

[11] Darwin, C. *Descent of Man,* New York: Barnes & Noble, reprint, 2004, p. 556.

[12] See for instance the letter Darwin wrote to W. Graham, July 3rd, 1881.

[13] Letter from Charles Darwin to George Darwin, his son, in 1873. In Himmelfarb, G., *Darwin and the Darwinian Revolution*, Chatto and Windus, 1958, p. 320

[14] Letter from Charles Darwin to Edward Aveling, Karl Marx's son in law around 1880. In Herbert, S., 'The Place of Man in the Development of Darwin's Theory of

Andrew Dixon White also revealed the effect that Lyell's capitulation to Darwinism had on the public mind, although any evidence of Lyell's machinations that White may have known about, were ignored. White's overall intention was incidentally to help establish scientism and a sense of conflict between science and Christian faith, and he praised Lyell for being an 'honest man.' With regard to this, White commented

> But in 1863 came an event which brought serious confusion to the theological camp: Sir Charles Lyell, the most eminent of living geologists, a man of deeply Christian feeling and of exceedingly cautious temper, who had opposed the evolution theory of Lamarck and declared his adherence to the idea of successive creations, then published his work on the *Antiquity of Man,* and in this and other utterances showed himself a complete though unwilling convert to the fundamental ideas of Darwin. The blow was serious in many ways, and especially so in two - first, as withdrawing all foundation in fact from the scriptural chronology, and secondly, as discrediting the creation theory. The blow was not unexpected; in various review articles against the Darwinian theory there had been appeals to Lyell, at times almost piteous, "not to flinch from the truths he had formerly proclaimed." But Lyell, like the honest man he was, yielded unreservedly to the mass of new proofs arrayed on the side of evolution against that of creation.[15]

Although Lyell had worked gradualism into geology he seemingly remained in favour of successive creations for many years, thus giving some comfort to those who believed in divine creation within an old earth framework. That is until after Darwin had published his work on evolution. Lyell then published his book *The Antiquity of Man* in 1863 and finally came to accept publicly Darwin's theory.

While White comments that Lyell was an 'honest man' in accepting the proofs of evolution, the evidence from his private letters, and confirmed by Darwin's correspondence, shows that Lyell and Darwin were working to a carefully hidden plan along the lines of the one Voltaire advocated. This was a 'slow' and 'silent' attack upon Christianity. The purpose of this was to achieve an outcome through concealment, but one that White presented as straightforward. It is questionable then whether

---

Transmutation: Part II,' *Journal of the History of Biology,* Vol. 10, no. 2, 1977, p. 161, In: Bowden, *The Rise of the Evolutionary Fraud,* p. 98

[15] White, A.D. *A History of the Warfare of Science with Theology in Christendom,* Chapter I, Part 4, 1898

Lyell was anything but honest, and along with Joseph Hooker was carefully involved with the oversight of *Origins of Species* from the beginning. Lyell and Hooker had read-out Darwin's paper, alongside one prepared by Alfred Russell Wallace, to the Linnaean Society of London on 1 July 1858. Interestingly, Lyell's religious beliefs were perhaps closer to Wallace than Darwin's. Lyell, as a Unitarian deist, believed that evolution had a guiding hand while Wallace had an interest in Spiritualism.

Paul Marston, an historian of geology, also argues that a number of leading geologists during the nineteenth century did not in fact fully accept gradualism in geology. The ascendancy of Lyell's version of geology was promoted towards the end of the nineteenth century, despite the scientific evidence and the academic view of geology that dissented from Lyell's gradualism. Marston identifies Archibald Geikie, with his book *Founders of Geology* published in 1897, as the one who developed Lyell's mythological status as the founder of modern geology within a uniformitarian setting. Marston suggests that the science of geology never fully accepted Lyell, but that the myth of gradualism is deeply entrenched in popular science nonetheless.[16] This entrenchment of the myth was seemingly necessary to support the theory of gradual, progressive evolution.

Furthermore, recent comments from palaeontologists and geologists reveal that there has been an acknowledgement to the way in which Lyell deliberately misrepresented the geological evidence. Even Stephen J. Gould for instance has commented that Lyell and Darwin did not prove gradualism from the rock layers, but instead imposed it upon the natural sciences through clever rhetoric. Gould goes further and says that their campaign had a largely negative impact on the science of geology by restricting fresh hypotheses and therefore closing minds within the profession. As such, alternatives to gradualism were ignored even though reasonable and empirically based.[17]

In an earlier paper written in 1975 Stephen J. Gould commented that Lyell's work was a brilliant brief for an advocate. Firstly, Gould noted, Lyell set up a strawman to destroy by misrepresenting the catastrophists, whereas in fact the catastrophists were the level headed empiricists of the day and not blinded by religious argumentation. Instead the appearance of the rocks showed them to be highly folded and faulted in places with whole faunas wiped out. Lyell instead used his imagination to overcome the literal appearance. Secondly, Lyell developed a false web of claims relating to

---

[16] Forster and Marston, *Reason, Science and Faith*, pp. 342-343
[17] Gould, S. J., 'Toward the vindication of punctuational change,' In: W. A. Berggren & J. A. Van Couvering (Eds.): *Catastrophes and earth History: The New Uniformitarianism*, Princeton N.J., Princeton University Press, 1984, pp. 14-16

uniformity by conflating methodological statements that must be accepted from first principles, with theoretical ones that were merely stated from prior commitments. As such his aim was to pull the wool over fellow scientist's eyes by forcing acceptance of his theoretical claims by hiding his prior propositions.[18] Derek Ager has also commented that geology got into the hands of theoreticians conditioned by the prevailing social and political history of the early nineteenth century. As a result modern geology allowed itself to be hoodwinked into avoiding interpretations of the past that involve catastrophic or extreme processes. Although both Ager and Gould have objected to creationists using their comments in this way, claims of malpractice cannot be suppressed.[19] However, despite recognition that catastrophes have played a part in shaping the earth amongst some scientists, the Lyell myth persists in many, more popular quarters. And the close link between Lyell and Darwin, and the secretive liberal agenda, must call into question the objectivity of Darwin's theory.

Chapter Summary

Following the publication of Darwin's work a number of high profile Christians remained opposed to the idea of evolution and were openly critical, even to the point of experiencing personal insults. Previously, the liberal agenda in the sciences had used clever rhetoric, persuasion, and charm to win the argument, but once the plan had blossomed the charm began to change to a culture of bullying. Through Darwin's personal correspondence Lyell's plan can be seen for what it is, a slow silent attack upon Christianity. But after publication the attacks became more openly vocal. There was also a revisionist programme that became evident, and this sought to establish the ascendancy of Lyell and Darwin's work to the point of mythology, for instance through the work of Geikie, and also a sense of conflict between science and Christian faith through the writing of Dixon White and Huxley.

---

[18] Gould, S.J. 'Catastrophes and the Steady State earth,' *Natural History*, Vol. LXXX, No. 2, Feb 1975
[19] Ager, D. V., *The Nature of the Stratigraphical Record*, The Macmillan Press Ltd, London, 1981, pp46-47

12.

Scientism and the Conflict Hypothesis

During the second half of the nineteenth century secular supporters of Darwinism encouraged the sense of conflict that arose between science and Christianity. This was in many ways an extension of the Whig scheme of Lyell, Darwin and Babbage to free the sciences from Moses. But there was more to it than that, as undoubtedly there had been conflict between conservative Anglicans and Enlightenment philosophers during the seventeenth and eighteenth centuries as well. What appears to have been new was an attempt at a revision of history in order to set the Church's perceived conservatism against the perceived beneficial progress of science as observed in history. Thomas Huxley was one of the main protagonists in this regard while in America Andrew Dixon White worked towards a similar goal. In public, Darwin presented an agnostic face, and at times seemed unsure whether his theory of gradual evolutionary change was correct. He was also apparently so concerned about the negative impact of his theory that he was unable to defend his work publicly at the Oxford debate in 1860 due to ill health. That task fell to Huxley, and amongst Darwin's followers he was perhaps the most determined in openly supporting evolution. Huxley was to gain the identity of 'Darwin's Bulldog,' and worked hard at promoting Darwin's theory.

Huxley and the X Club

Huxley was the most outspoken and prominent public supporter of Darwinian evolution and was seemingly determined in his desire to undermine organised Christianity, rejecting anything that he thought couldn't be explained by natural laws. With his self-confidence and bluster, he was a powerful speaker and this helped to cover up doubts he had over aspects of evolution through his life. He could not bring himself for instance to fully accept evolution as scientifically proved, and never fully rejected the possibility that God may exist either. However, Huxley helped to set up the secretive X Club with nine members in 1864. The stated aim of this newly formed organisation was to pursue scientific knowledge without any hindrance from what they saw as 'religious dogma,' although Christianity was of more concern than other beliefs that were prevalent at the time. Of less concern was interest in spiritualism and theosophy that was present in Victorian Britain, even amongst notable scientists such as evolutions co-founder, Alfred Wallace. Lyell was apparently Unitarian, while Ernst Haeckel was a pantheistic monist. Haeckel believed that a guiding hand had directed evolution, but not one acceptable to Christianity.

The nine members of the X Club were Thomas Huxley, Herbert Spencer, George Busk, Edward Frankland, Thomas Hirst, Joseph Hooker, John Lubbock, William Spottiswoode and John Tyndall. Six members were past Presidents of the British Association, while three were past Presidents of the Royal Society. Between them, they were able to control the direction of science and education policy in the Royal Society, and thus society as a whole. They met together on the evening before planned meetings of the Royal Society to discuss policy, and discussed other business such as the ongoing struggle against religion and the place of science in education.[1] In other words, the concerns of the X Club, and by extension the policy of the Royal Society, was to direct science and society away from Christian influence. While some members of the club seemed to have a measure of Christian concern, others such as Huxley were heavily involved in the promotion of a sense of conflict. He desired that science should be conducted without recourse to faith at all.

Huxley was concerned with the organisation of scientific knowledge, and wrote to Ernst Haeckel that he should remain in Germany and help in this endeavour. 'I would counsel you to stop at home, and as Goethe says, find your America here.... It is the organisation of knowledge rather than its increase which is wanted just now.'[2] Huxley also shared Lyell's desire to establish science education in terms of natural history, and believed that it is possible to see 'geological speculation' in terms of 'Evolutionism.' He wrote 'I conceive geology to be the history of the earth, in precisely the same sense as biology is the history of living beings.'[3]

One of his aims in developing science education was to challenge the apparent discipline and authority of the education system of the Roman Catholic Church that relied upon historical knowledge and continuity, but he also presented biology in the context of a comparative study of *forms* as opposed to a science of cause and effect.[4]

Huxley was passionate about the elevation of naturalistic science and it would seem that, like Darwin, Auguste Comte was a major influence

---

[1] Irvine, W., *Apes, Angels and Victorians*, London: Weidenfeld and Nicolson, 1956, p. 183

[2] Correspondence from Huxley to Haeckel, 7 June 1865

[3] Huxley, T.H., 'Geological Reform, (1869 Lay Sermons)', in *Collected Essays* Vol. VIII, London, 1893, pp. 316-317. See also Colling, A., *Science Matters: Discovering the Deep Oceans*, Open University, 1995, p. 31.

[4] Huxley, T.H., 'Scientific Education: Notes of an After-dinner Speech' (given in 1869 at the Liverpool Philomathic Society), in *Collected Essays*, Vol. III, London, pp. 120-124

on him. Although Huxley was critical of some of Comte's ideas, particularly Comte's desire to set himself up as a sort of high priest of science under a positivist religion, Huxley did however accept the positivist ideals. He at first commented that he was disappointed by Comte's religion of science, but then thanked Comte for suggesting that society should only be organised on a purely scientific basis.

> Great, however, was my perplexity, not to say disappointment, as I followed the progress of this "mighty son of earth" in his work of reconstruction. Undoubtedly "Dieu" disappeared, but the "Nouveau Grand-Être Suprême," a gigantic fetish, turned out brand-new by M. Comte's own hands, reigned in his stead. "Roi" also was not heard of; but, in his place, I found a minutely-defined social organization, which, if it ever came into practice, would exert a despotic authority such as no sultan has rivalled, and no Puritan presbytery, in its palmiest days, could hope to excel. While as for the "culte systématique de l'Humanité," I, in my blindness, could not distinguish it from sheer Popery, with M. Comte in the chair of St. Peter, and the names of most of the saints changed. ... Nothing can be clearer. Comte's ideal, as stated by himself, is Catholic organization without Catholic doctrine, or, in other words, Catholicism *minus* Christianity. ... Rightly or wrongly, this was the impression which, all those years ago, the study of M. Comte's works left on my mind, combined with the conviction, which I shall always be thankful to him for awakening in me, that the organization of society upon a new and purely scientific basis is not only practicable, but is the only political object much worth fighting for.[5]

Huxley at first rejected Comte's blueprint for society because it looked like 'Catholicism *minus* Christianity' with Comte as Pope. Comte's concept of a scientific priestly elite was also similar to Plato's ideal city-state where philosopher kings were to rule over society. There were also echoes with Bacon's blueprint for society involving a scientific judiciary. However, like Comte, Huxley also wanted science to form the only basis for the organisation of education, society and politics. Essentially then Huxley's argument was for the establishment of a non-conformist version of Comtism, effectively Comtism *minus* Comte. With his energy, and with his role with the X Club, Huxley and friends ensured that the Royal Society would maintain pre-eminence in science, and use it as an authority that

---

[5] Huxley, T.H., *The Scientific Aspects of Positivism, Lay Sermons, Addresses and Reviews*, London, 1870, pp. 129-133

could guide science without considering religious objections. While this may appear a neutral position, it was in effect elevating a philosophy of naturalism above Christian theism. In other words science was becoming atheistic instead of theistic. In this regard science became scientism and was elevated above religion, both in determining how scientific knowledge should be interpreted, and how it should impact upon society.

The increasingly godless Royal Society was promoted as a central authority in science. Although Huxley seemed to want to abandon all authority from science, it was really the religious authority he wanted exorcised most, and he was willing to work towards the establishment of a central controlling authority that was hostile to Christianity in the shape of the Royal Society. Huxley was a skilful rhetorician and used it to undermine Christian faith commenting that '...science, and the methods of science, are the masters of the world' and that 'quite frankly' it is

> ...almost beneath the dignity of my calling, as a man of science, to listen to [religiously motivated] objections as these. If it be *really* true that science is opposed to religion, all I can say is, so much the worse for religion. If science is *really* opposed to traditions, the sooner the traditions vanish and are no more seen or heard of, the better.[6]

Huxley also sought to place a higher burden of proof upon his opponents than he was willing to accept for himself. He noted for instance that '...the bringing into existence of an animal, at once, is a thing which is, in the nature of the case, capable of neither proof nor disproof...' As a result it is 'no subject for science', which can only be concerned with 'matters capable of proof or disproof.' Appealing to general natural laws Huxley then commented that he '...could not conceive but that these successive races *must* have proceeded from one another in the way of progressive modification' although stressing that it was only a hypothesis.[7] However, Huxley would only accept a rejection of this Darwinian hypothesis, which he already said was *beyond science*, on one based on naturalistic scientific explanations.

> ...if you see good to reject this hypothesis, if you think that my reasonings from the principles I started with are fallacious, or that those principles themselves are erroneous, reject it by all means; and if you can show me, *on these grounds*, that you are right, I will

---

[6] Huxley, T.H., 'Science and Religion,' *The Builder*, Vol. 17, Museum of Geology, January 1859, p. 35

[7] Huxley, *Science and Religion,* p. 35

reject it also as speedily as possible, and thank you for the refutation. Why should I encumber myself with the burden of an untruth?[8]

Huxley then placed the onus on his opponents to refute something he considered only a hypothesis and beyond science. This then begs the question why something that he considered beyond science needed scientific refutation at all. Incredibly though, as no one could scientifically refute something Huxley considered beyond science, he then considered it good and acceptable as science. Huxley cleverly shifted the burden of proof from himself to his opponents, but it was an invalid shift, and pure sophistry being devoid of logical coherence. In other words, Huxley's hypothesis had become set in his own mind as a scientific truth despite stating that a scientist could not prove it scientifically. For Huxley then, any causal argument taken from information in an historical religious text could have no place in science. Recall also that Darwin, Huxley and Lyell had effectively developed geology and evolution in historicist terms. But ironically they denied that an historical document could have any bearing on the matter if was the Christian Scriptures.

The idea that science is possible without the dependence of foundational worldview assumptions is really a fallacy. Comte was after all a philosopher not a scientist, and so in effect undermined his own claim by arguing that truth might arise through science alone. The later logical positivists also argued that only objectively verifiable statements could be proved true, but that statement is itself not objectively verifiable and thus it refutes itself. The philosophy of Comte, Huxley and of the later logical positivists is therefore in fact self-refuting.

But Huxley's confidence blinded him to his errors. In effect, he accepted his own opinion as truth even though he admitted he could not prove it scientifically. It may be noted as well that Huxley made a philosophical leap in his reasoning, applying the apparent laws of the universe to the laws of the evolution, including of mankind, in a complete cosmic process. Here again Huxley's own naturalistic philosophy is acceptable as science, while Huxley dismissed religiously based cosmology as non-science. In his own words, '... science, and the methods of science, are the masters of the world.' In the next couple of paragraphs, Huxley continued to engage in sophistry in his attacks against Christian faith. On the one hand he stated that it is wrong to claim that there is antagonism between science and religion, but then carefully defined science as a

---

[8] Huxley, *Science and Religion,* p. 35

religious practice, and stated that religion only has value if it is treated scientifically.

> But it is not true. If you have seen occasion to put any faith in what I tell you, believe me now when I say, that of all the miserable superstitions which have ever tended to vex and enslave mankind, this notion of the antagonism of science and religion is the most mischievous.

> True science and true religion are twin-sisters, and the separation of either from the other is sure to prove the death of both. Science prospers exactly in proportion as it is religious; and religion flourishes in exact proportion to the scientific depth and firmness of its basis.[9]

Science therefore may take its place as a religion, and true religion, which is faith, has no value because true religion requires a scientific base, not a faith base. There is irony in this because Huxley, having condemned scientific arguments that are motivated by religion, then asserts that science too should be treated as a religion. However, while Huxley was making bold statements that the scientific community was the new 'truth seeker' fighting off the hindrance of organised Christianity, he himself became enmeshed in an unfortunate episode regarding a search for the protoplasm of life. Such episodes became surprisingly commonplace in the twentieth century as pressure grew to find evidence for evolution, sometimes through over enthusiasm, at other times through deliberate malpractice. Sometimes scientists were so keen to find evidence that they made basic mistakes, at other times the scientists engaged in fraud in support of their goals. Darwin had argued in *Origins* that the transitional fossils would be found if sufficient effort was put into the search, and from the second half of the nineteenth century onwards, geology was given over to the search for these missing links.

Response from the Victoria Institute

In response to the rhetoric and hegemony of Huxley and other supporters of Darwin a group of Christian scientists formed the Victoria Institute, or Philosophical Society of Great Britain in 1865. The purpose was to mount an effective challenge to attacks on Scripture, as their concern was to counter the growing anti-Christian bias that they believed

---

[9] Huxley, 'Science and Religion', p. 35

was evident within science and society. The founding members of the Victoria Institute considered that pseudo-science was in effect replacing real science and then used to deliberately undermine Scripture. Untested theories, they believed, were developed, promoted and accepted without the necessary supporting evidence usually expected by the scientific community. They were also concerned with apparent compromise by some leading clerics because it opened the door to atheists and deists. In *Scientia Scientiarum*, the author referred to correspondence between Adam Sedgwick and William Cockburn, Dean of York.[10] In 1844 Cockburn had attacked the nebular theory for the origin of the earth supported by Dean William Buckland in *The Bridgewater Treatises*. *The Bridgewater Treatises* were supposed to offer support to Scripture, but many felt there was too much accommodation in Buckland's writing. Dean Cockburn instead argued that no geological facts exist to support the long ages of the world. Although some nineteenth century scientists used this nebular theory to undermine the Mosaic account, it was later rejected when it became clear that volcanic rocks were often laid down in the presence of water. But such rejection of a failed hypothesis did not mean that the associated rejection of Scripture was revised. Failed naturalistic or uniformitarian hypotheses are soon replaced by others; this because the science of origins is driven by an overriding commitment to metaphysical naturalism. Naturalism thus rejects the Scriptural account as an explanation even before science begins. Reddie used this example of the nebular hypothesis to highlight how scientific theories, that are used to dismiss Scripture, are later abandoned as new evidence becomes known, but Scripture is never allowed a foot back inside the door of science. The rules of the game of science are stacked against Scripture from the start because of a commitment to the philosophy of naturalism. Thus defending Scripture against false science such as this formed the basis for the purposes of the Victoria Institute. The first Object of the Victoria Institute read

> *To investigate fully and impartially the most important questions of Philosophy and Science, but more especially those that bear upon the great truths revealed in Holy Scripture, with the view of defending these truths against the oppositions of Science, falsely so called.*

It is apparent that the original aims and purposes of the Victoria Institute were very similar to the aims and purposes of modern day creationist groups. Later the Victoria Institute increasingly moved away

---

[10] Probably James Reddie, the Honorary Secretary of the Victoria Institute

from literal readings of Scripture, and in the early 1930s leading members, including the President Sir Ambrose Fleming, formed a breakaway movement. This was called, the Evolution Protest Movement, now the Creation Science Movement.

A Growing Sense of Conflict

While Huxley was engaged in rhetoric against Christianity in Britain, in America two philosophers and historians of science developed the conflict hypothesis further. The first to publish was John William Draper who wrote the *History of the Conflict between Religion and Science* in 1874. This work targeted specifically the Roman Catholic Church. He argued in the preface that the

> ...history of Science is not a mere record of isolated discoveries; it is a narrative of the conflict of two contending powers, the expansive force of the human intellect on one side, and the compression arising from traditionary faith and human interests on the other.[11]

In other words it sets human progress against religious beliefs and values. Dixon White, a student of Yale and member of the Skull and Bones fraternity, became the first President of Cornell University and developed the warfare thesis between science and faith over a period of around thirty years. White was preparing the university to develop science without religious influence, announcing upon its foundation in 1865 that it would be an 'asylum for Science – where truth shall be sought for truth's sake, not stretched or cut exactly to fit Revealed Religion.'[12]

White further developed his thought in 1869 in a lecture entitled *The Battle-Fields of Science* and later in 1896 publishing *History of the Warfare of Science with Theology in Christendom*. It was in effect a revision of history to suit the mythology of gradualism and evolution. He argued that science eventually triumphed over Flood geology when in 1863 Charles Lyell published a final rejection of the Deluge in *The Antiquity of Man*.[13] He also argued that Columbus, in seeking a western passage to the

---

[11] Draper, J.W. *History of the Conflict Religion,* New York: D. Appleton and Co. 1881

[12] In: Lindberg, D.C., & Numbers R.L., (eds) 'Introduction,' in *God & Nature: Historical Essays on the Encounter between Christianity and Science*, Berkeley: University of California Press, 1986, pp. 2-3

[13] White, A.D., *A History of the Warfare of Science with Theology in Christendom*, (Chapter V, Part IV) New York: D. Appleton and Co. 1896

Indian continent, was fighting against the Church authorities and ignorant superstition. Such ignorance, White thought, included a belief that the earth was flat and that sailors who ventured too far from the land would fall off the edge.[14] However, a much more complicated picture emerges in reality concerning historical beliefs about the size and shape of the earth, and White and Draper's thesis of conflict has been widely, and rightly criticised.

White and the Onondaga Giant

White also seemed to have had a strong interest in the discovery of the Cardiff or Onondaga Giant in the autumn of 1869. He was notably present to warn scientists about the error, and to observe the associated human folly of the common people. At face value White seems to have only been a casual observer of the events, but he noted people's behaviour, and carefully warned scientists to avoid getting caught up in the excitement. The Cardiff Giant was a nine-foot long statue of a man buried on a local farm. It was placed five feet below the ground level in a large pit and apparently made of the local grey Onandaga limestone. White commented that

> ...it was a stone giant, with massive features, the whole body nude, the limbs contracted as if in agony, together with the appearance of 'great age' due to deep grooves and channels in its under side, apparently worn by the water which flowed in streams through the earth and along the rock on which the figure rested.[15]

White noted that there was a degree of quiet solemnity surrounding the place with visitors speaking in whispered tones. Of course he claimed that he had believed it to be a hoax at once, but found it 'especially interesting' to observe the 'evolution of myth and legend' from the local Christians who he claimed found a 'joy in believing.' Within a week White found that one story had developed that the Onondaga statue was the petrified body of a gigantic Indian prophet who had flourished many centuries ago. But in addition to this there were suggestions from others who seemingly found it a confirmation of the biblical text that there were 'giants in those days' (from *Genesis 6:4*).[16]

---

[14] White, 1896, *A History of the Warfare of Science with Theology in Christendom*, (Chapter II, Parts III, IV and V)

[15] White, A.D. *Autobiography of Andrew Dixon White*, Vol. II, New York, The Century Co. 1917, pp. 465-485

[16] White, A.D. *Autobiography,* pp. 465-485

Another prominent clergyman who came to the find said, according to White, that it was 'not a thing contrived of man, but is the face of one who lived on the earth, the very image and child of God.' However, while White was carefully observing the Christians, he was careful to advise scientists against the find. For instance, a number of scientists who came to Syracuse met with White in a hotel where he warned them of the nature of the statue. He urged them to be especially cautious as a mistake might prove damaging to their reputation and to the proper standing of science and scientific men if the matter were to prove a fraud.[17] Indeed, it was a fraud carried out by the farmer William C. Newall and his brother in law George Hull. It would seem that Hull, who was an atheist, devised the plan to mock the faith of believers in Genesis regarding giants mentioned in the Bible. This plan arose following an argument with the Reverend Turk of Ackley Iowa. Both Newall and Hull incidentally made a financial profit from their work in selling tickets and the statue. White typically used this episode as part of his conflict hypothesis against Christians in science, and it is notable that he only seemed cautious to warn scientists, while observing the activities of the general people. Most of those people in fact thought it was an ancient Indian idol or statue.

Chapter Summary

While the Cardiff Giant provides an interesting story, what is evident more generally is that there was a determined attempt at creating a sense of conflict between Christianity and science during the nineteenth century, especially during the second part of that century. Towards this end, a great deal of rhetoric was used to make up for lack of coherence, both in Britain and in America. The X Club met to control the agenda of the Royal Society, and Thomas Huxley promoted evolution heavily because of a dislike of the Christian Church, even though he secretly had doubts about Darwin's theory. Biology and geology were also set in the context of history and a comparative study of Platonic *forms*; an historical *narrative* imposed upon science in opposition to Christianity. Draper and Dixon White argued that historically there had been a conflict between science and faith, and revised history to suit the needs of the new science. The fact that the evidence does not fit such a hypothesis was of little concern to them. In fact, as shown in earlier chapters, there is evidence that biblical literalism has instead advanced science.

---

[17] White, A.D. *Autobiography,* pp. 465-485

13.

Bathybius haeckelii and a Reign of Terror

Thomas Huxley and Ernst Haeckel were keen to find evidence for the missing links, that Darwin proposed existed, in the second half of the nineteenth century. One of the most important was the primeval protoplasm that was believed to be the simplest and earliest living organism.

It was Huxley who brought *Bathybius* into the world. There was initial excitement, along with support and encouragement from Haeckel. Haeckel believed strongly that life had naturally arisen from non-life in the past. However, this 'discovery' was later to cause a lot of embarrassment to these two men. The accusation was that their judgment was clouded due to their affinity for the Darwinian presuppositions. Huxley, had of course spent a period at sea as a marine naturalist. From this experience he was able to examine preserved samples of deep sea floor sediment that had been collected by the H.M.S. *Cyclops* in 1857; at the time it was engaged in laying trans-Atlantic telegraph cables. Huxley was employed on the ship to study the sediment and benthic organisms from the sea floor.

One sample of muddy sediment, preserved in alcohol, appeared to move and a have life of its own. A thin 'jelly-like' film of mucus had settled on the surface with the appearance of tiny granules. His excitement was raised further through microscopic studies, and this led to the claim that he had found the original protoplasm of life. Haeckel had recently proposed the existence of such a living organism, and such a discovery suited the Darwinian speculation of the time. However, in his excitement he did not show proper scientific caution, and carry out the necessary chemical tests. Instead he launched into promotion of this find with his usual bluster and confidence as the most basis form of life. In a letter to Haeckel in 1868 he claimed that it was

> ...a new "Moner" which lies at the bottom of the Atlantic to all appearances, and gives rise to some wonderful calcified bodies. I have christened it *Bathybius Haeckelii,* [sic] and I hope you will not be ashamed of your god-child. I will send you some of the mud with the paper.[1]

---

[1] Huxley. T.H., *Letter to Ernst Haeckel*, 6th October 1868

Bathybius haeckelii

Huxley named this apparent single celled organism after his German friend Haeckel, as *Bathybius haeckelii*. Haeckel was at this time the main protagonist for *abiogenesis*; the hypothesis that living organisms might arise out of non-living chemicals all by themselves. Thanks to Huxley he was able to examine *Bathybius* for himself, and concluded with enthusiasm that it was the original primordial slime (*Urschleim* in German). Huxley commented in a paper in *Nature* in 1870 that

> The longest of the papers ... is devoted to a careful study of *Bathybius,* and the associated Coccoliths and Coccospheres; and it is a mattaer [sic] of great satisfaction that Prof. Haeckel has arrived at conclusions which, in all the main points, agrees with my own respecting these remarkable organisms.[2]

Huxley then was influenced by comparison with coccoliths, and also penicillin. Haeckel praised Huxley for this work, and the extension of the theory of evolution to a universal progressive principle; from the lowest form of life to mankind. Haeckel wrote a glowing endorsement in *Nature* praising Huxley for his contribution to Darwin's evolution; entitled 'Scientific Worthies: Thomas Henry Huxley.'

> After Charles Darwin had, in 1859, reconstructed this most important biological theory, and by his epoch-making theory of Natural Selection placed it on an entirely new foundation, Huxley was the first who extended it to man, and in 1863, in his celebrated three Lectures on "Man's Place in Nature," admirably worked out its most important developments. With luminous clearness, and convincing certainty, he has here established the fundamental law, that, in every respect, the anatomical differences between man and the highest apes are of less value than those between the highest and the lowest apes. Especially weighty is the evidence adduced for this law, in the most important of all organs, the brain; and by this, the objections of Prof. Richard Owen are, at the same time, thoroughly refuted. Not only has the Evolution Theory received from Prof. Huxley a complete demonstration of its immense importance, not only has it been largely advanced by his valuable comparative researches, but its spread among the general public has been largely due to his well-known popular writings. In these

---

[2] Huxley. T.H., 'Life in the Deep Sea,' *Nature,* 2nd July 1870, p. 187

he has accomplished the difficult task of rendering most fully and clearly intelligible, to an educated public of very various ranks, the highest problems of philosophical Biology. From the lowest to the highest organisms, from Bathybius up to man, he has elucidated the connecting law of development.[3]

Haeckel and Huxley were in the process of developing a unified and complete evolutionary theory. This was to be a grand progression from the simplest forms, to vertebrates, mammals and finally to human beings. *Bathybius* was apparently central to this process, and pressing upon the minds of those promoting Darwinism. Following the *Cyclops* work, two other expeditions were commissioned, one on H.M.S. *Lightning,* the other on H.M.S. *Porcupine,* but these voyages failed to find fresh samples of *Bathybius.* However, the *Challenger* expedition of 1872 had greater success as the scientists aboard noted something unusual. Samples of sea floor sediment preserved in alcohol showed evidence of *Bathybius,* although it was noted that samples of sediment stored in seawater did not. John Buchanan, the ship's chemist, tested the jelly-like substance, and found that it was in fact hydrated calcium sulphate ($CaSO_4.2H_2O$). The jelly mucus was caused by a reaction of the alcohol on the mud. Upon hearing the news, Huxley realised that a mistake had been made. In response he wrote to the journal *Nature* reporting the fact that *Bathybius* was probably not organic.

> Prof. Wyville Thomson further informs me that the best effort of the Challenger's staff have failed to discover *Bathybius* in a fresh state, and that it is seriously suspected that the thing to which I gave that name is little more than sulphate of lime, precipitated in a flocculent state from the sea-water by the strong alcohol in which the specimens of the deep-sea soundings which I examined were preserved.[4]

In further correspondence with Michael Foster, Huxley commented that.

> I have just had a long letter from Wyville Thomson. The *Challenger* inclines to think that *Bathybius* is a mineral precipitate! In which case some enemy will probably say that it is a product of my precipitation. So mind, I was the first to make that "goak." Old Ehrenberg suggested something of the kind to me, but I have not

---

[3] Haeckel, E., 'Scientific Worthies: Thomas Henry Huxley,' *Nature*, Feb. 1874, p.258
[4] Huxley. T.H., 'Notes from the Challenger,' *Nature,* **12**, Aug. 1875, pp.315-316

his letter here. I shall eat my leek handsomely, if any eating has to be done.[5]

In 1875 then Huxley appeared to accept that *Bathybius* was an illusion of science; a mistake. However, he continued to hold open a measure of doubt, even in 1879 in this regard. He wrote that the question of his 'friend' was still not settled, stating that 'my own judgment is in an absolute state of suspension about it.'[6] Perhaps the reason for this equivocation was to do with saving the face of the President of the British Association for the Advancement of Science (BAAS) who had continued to offer support to *Bathybius* at the annual meeting. Science historian Angela Colling has recently commented that *Bathybius* played an important role in offering support to the theory of evolution. This meant that many scientists wanted it to be true, and therefore it clouded their judgment, and this included the Huxley and Haeckel .[7]

The Duke of Argyll, George Campbell, later criticised the scientific community for allowing itself to be caught up in a wave of emotional attachment that blinded its critical faculties. Argyll wrote in 'A Great Lesson' in 1887 that

> The naturalists of the 'Challenger' began their voyage in the full Bathybian faith. But the sturdy mind of Mr. John Murray kept its balance–all the more easily since he never could himself find or see any trace of this pelagic protoplasm when the dredges of the 'Challenger' came fresh from bathysmal bottoms. Again and again he looked for it, but never could he discover it. It always hailed from home. The bottles sent there were reported to yield it in abuirdance,[sic] but somehow it seemed to be hatched in them. The laboratory in Jermyn Street was its unfailing source, and the great observer there was its only sponsor. The ocean never yielded it until it had been bottled. At last, one day on board the 'Challenger' an accident revealed the mystery. One of Mr. Murray's assistants poured a large quantity of spirits of wine into a bottle containing some pure sea-water, when lo! the wonderful protoplasm Bathybius appeared. It was the chemical precipitate of sulphate of lime produced by the mixture of alcohol and sea-water. This was bathos indeed. On this announcement 'Bathybius' disappeared from

[5] Huxley. T.H., *Letter to Michael Foster*, 11th August 1875
[6] Huxley. T.H., 'Report to BAAS,' *Nature*, 28th August 1879
[7] Colling, A., *Science Matters: Discovering the Deep Oceans*, Open University, 1995, p. 29

science, reading us, in more senses than one, a great lesson on 'precipitation.'[8]

The Duke pointed out that *Bathybius* was just a 'slimy mucus,' and 'structureless to all microscopic examination.' As such it was similar to a lot of other sedimentary material found on the sea floor. The level of enthusiasm blinded some of the leading scientists, or perhaps there was a deliberate conspiracy. The Duke speculated that.

> Here was a grand idea. It would be well to find missing links; but it would be better to find the primordial pabulum out of which all living things had come. The ultra-Darwinian enthusiasts were enchanted. Haeckel clapped his bands and shouted out Eureka loudly. Even the cautious and discriminating mind of Professor Huxley was caught by this new and grand generalisation of the 'physical basis of life.' It 'was announced by him to the British Association in 1868. Dr. Will. Carpenter took up the chorus. He spoke of 'a living expanse of protoplasmic substance,' penetrating with its living substance the 'whole mass' of the oceanic mud.' A fine new Greek name was devised for this mother slime, and it was christened 'Bathybius,' from the consecrated deeps in which it lay. The conception ran like wildfire through the popular literature of science, and here again there was something like a coming Plebiscite in its favour. Expectant imagination soon played its part. Wonderful movements were seen in this mysterious slime. It became an 'irregular network,' and it could be seen gradually 'altering its form,' so that 'entangled granules gradually changed their relative positions.

> This is a case in which a ridiculous error and a ridiculous credulity were the direct results of theoretical preconceptions. Bathybius was accepted because of its supposed harmony with Darwin's speculations.[9]

The lack of care shown by Huxley in this affair is highlighted by his own statement, where; 'The man of science, in fact, simply uses with scrupulous exactness the methods which we all, habitually and at every

---

[8] Duke of Argyll, 'A Great Lesson,' *The Nineteenth Century* **22,** Sept. 1887, pp.293-309

[9] Duke of Argyll, 'A Great Lesson,' pp.308-309

moment, use carelessly.'[10] His careless mistake did however serve the purpose of promoting evolution as a truth of science for at least seven years.

Was this error just a simple mistake? One may wonder whether such carelessness correlates with Huxley's position as a leading scientist, but others were taken in by the desire to find evidence in support of Darwinism. Huxley of course protested that he was innocent of the charge of deliberate malpractice. However, we may note, from his own words and those of Haeckel, that *Bathybius* was an important part of the project to promote Darwin's theory. Haeckel had commented that 'From the lowest to the highest organisms, from Bathybius up to man, he [Huxley] has elucidated the connecting law of development,'[11] and Huxley said that 'Haeckel has arrived at conclusions which, in all the main points, agrees with my own respecting these remarkable organisms.'[12]

Reign of Terror

The Duke of Argyll, a former president of the British Association, later complained that there existed a *Reign of Terror* in the scientific community against anyone who disagreed with Darwin's position. Evidence of a campaign of bullying against those who disagreed with Darwinian evolution was growing. However, the cause of this new accusation was related to findings regarding the formation of coral. The Duke had commented on Huxley's use of personal attacks against John Murray. In arguing his case he commented that 'My sincere respect for Professor Huxley forbids me from following him into the field of personal polemics, even if this Review were a fitting place for such exercitations.'[13]

The marine scientist Murray was put under pressure not to publish his research that found evidence that contradicted Darwin's hypothesis. Darwin had followed Lyell's belief that coral grew slowly on volcanic rocks as an ancient continent slowly subsided into the Pacific Ocean. They did not think that coral could grow on soft sediment. Murray however showed that softer sediment could provide an adequate surface for coral growth, and that a reef may grow upwards towards the sea surface. The Duke of Argyll commented that

---

[10] Huxley, T.H. *On the Education Value of the Natural History Sciences*, 1854, in Colling, A., *Science Matters: Discovering the Deep Oceans*, p.32-33
[11] Haeckel, 'Scientific Worthies: Thomas Henry Huxley,' 1874
[12] Huxley, 'Life in the Deep Sea,' 1870
[13] Duke of Argyll, 'A Reply: Science Falsely So Called,' *The Nineteenth Century* **21,** May 1887, pp.771-774

In a recent article in this Review I had occasion to refer to the curious power which is sometimes exercised on behalf of certain accepted opinions, or of some reputed Prophet, in establishing a sort of Reign of Terror in their own behalf, sometimes in philosophy, sometimes in science. This observation was received I expected it to be–by those who being themselves subject to this kind of terror are wholly unconscious of the subjection. It is a remarkable illustration of this phenomenon that Mr. John Murray was strongly advised against the publication of his views in derogation of Darwin's long-accepted theory of the coral islands, and was actually induced to delay it for two years. Yet the late Sir Wyville Thomson, who was at the head of the naturalists of the 'Challenger' expedition, was himself convinced by Mr. Murray's reasoning, and the short but clear abstract of it in the second volume of the Narrative of the Voyage has since had the assent of all his colleagues.

It is that Darwin's theory [of coral formation] is a dream. It is not only unsound, but it is in many respects directly the reverse of truth. With all his conscientiousness, with all his caution, with all his powers of observation, Darwin in this matter fell into errors as profound as the abysses of the Pacific. All the acclamations with which it was received were as the shouts of an ignorant mob. It is well to know that the plebiscites of science may be as dangerous and as hollow as [t]hose of politics. The overthrow of Darwin's speculation [of coral formation] is only beginning to be known. It has been whispered for some time. The cherished dogma has been dropping very slowly out of sight. Can it be possible that Darwin was wrong?[14]

Huxley of course denied that there existed a *Reign of Terror* against those who rejected Darwinism. In private correspondence Huxley wrote that the Duke had been causing mischief over the *Bathybius* affair. He wrote that '…the theologians cannot get it out of their heads, that as they have creeds, to which they must stick at all hazards, so have the men of science. There is no more ridiculous delusion.'[15] But even in 1890 Huxley was still complaining that 'Bathybius is too convenient a stick to beat this dog with to be ever given up…'[16] But Huxley was in denial, writing a few years earlier in 1887 the following.

---

[14] Duke of Argyll, 'A Great Lesson,' 1887
[15] Huxley. T.H., *[Letter to unknown person]*, 30th September 1887
[16] Huxley. T.H., *Letter to John Donnelly*, 10th October 1890

What is meant by my being caught by a generalization about the physical basis of life I do not know; still less can I understand the assertion that *Bathybius* was accepted because of its supposed harmony with Darwin's speculations. That which interested me in the matter was the apparent analogy of *Bathybius* with other well-known forms of lower life, such as the plasmodia of the Myxomycetes and the Rhizopods. Speculative hopes or fears had nothing to do with the matter; and if *Bathybius* were brought up alive from the bottom of the Atlantic to-morrow the fact would not have the slightest bearing, that I can discern, upon Mr. Darwin's speculations, or upon any of the disputed problems of biology. It would merely be one elementary organism the more added to the thousands already known.[17]

Huxley and Haeckel had of course already expressed in writing the importance of finding the original protoplasm of life. The Duke of Argyll pressed home his claim that a commitment to Darwinian presuppositions led to false claims of evidence and bullying. In *Science falsely so called*, the Duke later commented that Huxley had a habit of moving from evidential science to theory and metaphysics without acknowledging the switch in reasoning, and made false appeals to authority.

The first of these [points] concerns the use which Professor Huxley makes of the word 'science.' In common parlance this word is now very much confined to the physical sciences, some of which may be called specially experimental sciences, such as chemistry, and others exact sciences, such as astronomy. But Professor Huxley evidently uses it in that wider sense in which it includes metaphysics and philosophy. Under cover of this wide sweep of his net, he assumes to speak with the special authority of a scientific expert upon questions respecting which no such authority exists either in him or in anyone else. It seems to be on the strength of this assumption that he designates as pseudo-science any opinion, or teaching, or belief, different from his own.[18]

The Duke suggested that Huxley used elaborate arguments to blind his audience; an example relates to the vertebral origin of the skull. Huxley

---

[17] Huxley, T.H., 'An Episcopal Trilogy, Science and the Bishops,' *The Nineteenth Century,* 22, pp.625-40, November 1887, In: *Essays on Some Controverted Questions*; *Collected Essays V*: pp.126-59

[18] Duke of Argyll, 'A Reply: Science Falsely So Called,' 1887

dismissed a relatively minor point with clever rhetoric, and then developed sweeping statements in philosophy relating to a unity of organisation to support his case, but failed to provide sufficient evidence.

> I will illustrate what I mean by an example. One of the most elaborate of Professor Huxley's own works is his volume on *The Elements of Comparative Anatomy,* published some twenty-three years ago. Comparative anatomy is one of the branches of the larger science of Biology in which Professor Huxley is an expert; and, like all the other branches which grow out of the one great stem of 'Life,' as a subject of physical investigation, it runs up into ideas and conceptions which belong to, or border on, the region of metaphysics. In that volume Professor Huxley deals with the well-known question of comparative anatomy whether the vertebrate skull can, or cannot, be 'interpreted' as a developed vertebra. Through an elaborate argument, strictly conducted on the observation and analysis of physical facts, Professor Huxley comes to the conclusion that this 'interpretation' breaks down. 'The vertebral hypothesis of the skull,' he says, 'seems to me to be altogether abolished.' Yet, whilst rejecting this particular 'interpretation,' he accepts and enforces the general conception that there is a complete 'unity of organisation' between all vertebrate skulls, from the skull of a man down to the skull of a pike. Furthermore, Professor Huxley explains that by this 'unity of organisation' he means that all vertebrate skulls 'are organised upon a common plan.' Repeating the same idea in another place, he says, 'osseous skulls are constructed upon a uniform plan.'[19]

The reason why the Duke highlighted this sort of thing is that Huxley was in the habit of intimidating people on the basis of his scientific authority, even though such authority was lacking. He commented that

> I have dwelt upon this point because men are very apt to be intimidated by authorities in 'science,' when in reality no sort of authority exists. Professor Huxley talks about 'intellectual sins' quite in the language and spirit of the Vatican. I know a good many scientific men of the very highest standing who totally dissent from Professor Huxley's metaphysics and philosophy; and are by no means inclined to accept his expositions, even of physical science,

---

[19] Duke of Argyll, 'A Reply: Science Falsely So Called,' 1887

when those expositions travel beyond the particular branch in which he is an original observer.[20]

The Duke noted that new doctrines in science sometimes descend into a sort of fanaticism. However, in the case of evolution this development was more pronounced than normal, and he questions Huxley's own comments in this regard.

In conclusion, let me express a hope that Professor Huxley will yet do an important service to science, by entering in some detail upon a subject to which I have only alluded in passing, but in terms which have excited his astonishment. He says, most truly, that 'as is the case with all new doctrines, so with evolution, the enthusiasm of advocates has sometimes tended to degenerate into fanaticism, and mere speculation has, at times, threatened to shoot beyond its legitimate bounds.' These words indicate vaguely and tenderly, but significantly, a fact which I stated, and will again state with emphasis. There has been not merely a tendency to degeneration into fanaticism, but a pronounced development of it, and a widespread infection from it in the language of science. But it will be enough if Professor Huxley will explain fully what he means by this 'tendency,' and if he will specify wherein it has been shown. This is a work which has yet to be done. The knowledge of a great expert would help Professor Huxley to do it sooner than it could be done by others. They can only work with the materials which are supplied by such as he. It is a work which has begun, and which his own warnings have encouraged. Since he has authority to deal with 'intellectual sins,' let him convict, and lay bare, and anathemise this one which he treats so gently. The tendency of new doctrines to degenerate into fanaticism is one of the 'laws' to be traced in the long history of human follies, and all those who help to resist it are among the benefactors of their kind. I trust Professor Huxley may yet be with us for many years to come, and that he may expand and emphasise the hints and warnings he has given.[21]

Chapter Summary

The episode of *Bathybius* exposes Huxley and Haeckel to the claim that their own presuppositions clouded their judgment, to the point that they

---

[20] Duke of Argyll, 'A Reply: Science Falsely So Called,' 1887
[21] Duke of Argyll, 'A Reply: Science Falsely So Called,' 1887

did not conduct scientific research in a truly objective manner. At best it would seem that they were misled by Darwinian preconceptions; self-deception if you like. Despite claims to the contrary, their own correspondence and writing reveals the importance of finding something like *Bathybius* in order to help establish the Darwinian claims.

What may be seen with this affair, and other examples, is that a pattern emerges. Flimsy evidence for evolution is found and then used to promote the theory with an inappropriate level of certainty and confidence. Those who raise objections face personal attacks, insults and ridicule, and sometimes lose their careers and positions, even to this day. There is great pressure to conform through fear of a *reign of terror*. And the script of the *Bathybius* affair has been repeated through history, whether it is a pig's tooth in Nebraska, or odd artefacts in Piltdown. Deliberate attempts at deception have been used in some instances, but later exposed as fraudulent. Or too great a level of certainty has been attached to flimsy evidence.

14.

Dating the Earth and Plate Tectonics

As discussed, the French thinkers were among the first to try and estimate the age of the earth in the eighteenth century. While some of the ancient Greek philosophers had considered the earth to be eternal and cyclical, de Maillet instead relied upon Hindu texts to claim an age of the earth extending back two billion years. The Hindu scheme was incidentally also cyclical. The other notable French philosopher who attempted to date the earth was Comte de Buffon who arrived at a figure of 75,000 years based on the rate at which the earth might have cooled. Later this figure was extended to ten million years or more in private correspondence. Hutton and Lyell also considered the earth to be ancient, but they did not seek to date the planet, perhaps recognizing the difficulties they would have faced, although they believed that there had been a beginning. Charles Darwin on the other hand toyed with attempts at dating the earth, although he found that this caused him considerable trouble. In the first edition of *Origins*, Darwin ventured to make a crude estimation for the age of a particular valley in Southern England, noting that it would have taken 300 million years for erosion to denude the Weald from current estimates of water flow. This was long enough, he thought, for natural selection to fulfill its task, and usefully corresponded with Erasmus Darwin's belief in 'millions of ages.'[1] The problem for Darwin though was that this estimation left him open to attack from his many critics and it was removed by the time of the third edition of his work.

Lord Kelvin and the Age of the Earth

William Thompson, known as Lord Kelvin, had been working on a method of dating the earth while Darwin was writing his book, and first announced his estimation in 1862 to the Edinburgh Royal Society in a paper entitled *On the Secular Cooling of the earth.*[2] Kelvin had begun the task in 1844 and returned to the methodology of Buffon, where he thought

---

[1] Darwin, C., *On the Origin of Species*, First Edition, 1859, pp. 282-287, and Erasmus Darwin's *Zoonomia*, (in three parts), 1803, Vol. 1, p. 397.

[2] Also in *North British Review*, W. Thompson (1862) pp. 391-392. Darwin criticised Thompson's work in a letter to Charles Kingsley (10 June 1867) on the basis that there was wide disagreement on dates between Thompson, Samuel Haughton and William Hopkins. Later in 1864 Haughton determined a longer period of some 2 billion years to the beginning of the Tertiary period from the formation of the primordial ocean.

the age might be determined by estimating the rate of cooling from a once very hot body. With renewed vigour he continued his quest in December 1860 specifically to challenge the claims of Darwin.

Gravity was considered the source of energy in the sun, where the inward pressure was responsible for generating and releasing heat, but this could not last forever. From this assumption Kelvin arrived at a figure of between 20 million and 400 million years, suggesting an average figure of 100 million years, and he criticised Darwin's estimation of 300 million years for the formation of the English valley. He also believed in the meteoric theory, where it was believed that meteorites had fallen into the sun in the primordial solar system. This would have extended the age of the sun substantially beyond that given by the chemical theory alone, which was considered far too low. The chemical theory, he thought, would provide only enough energy for 3,000 years of heat. However, Darwin was skeptical of the divergence of estimates in Kelvin's work. Through the following decades Kelvin refined his work, and by the end of the 1880s his estimation moved towards the lower figure of 20 million for the age of the earth, an age far too small for Darwin's theory of natural selection to be accepted as fact. Darwin believed that his theory required at least several hundred million years, and saw Kelvin's work as a real thorn in the flesh as an 'odious spectre' that greatly hindered acceptance of his theory.[3] Such estimations were profoundly depressing for him and even at the end of his life he believed the earth would prove much older.

Huxley was also publicly critical of Kelvin's work, but failed to provide an objective response; instead he resorted to typical rhetorical argumentation. Huxley commented in the Presidential Address to the Geological Society in 1869 that although he didn't wish to call into question the accuracy of such distinguished mathematicians as Kelvin, he believed the work in dating the solar system to be an occasion when an error had been made. He commented that '...the admitted accuracy of mathematical processes is allowed to throw a wholly inadmissible appearance of authority over the results obtained by them.'[4] Ironically, Huxley and others have been far less concerned about false claims to authority in science when long ages supported the Darwinian hypothesis.

Dating in the Twentieth Century

It wasn't until the early twentieth century that new developments in science enabled the supporters of Darwin to return to the task of estimating the age of the earth, thus extending it further in support of a purely

---

[3] Bibby, C., *Scientist Extraordinary: T.H.Huxley*, Pergamon 1972, p. 49
[4] Bibby, *Scientist Extraordinary*, p. 50

naturalistic account of origins. The problems associated with Lord Kelvin's estimation were then sidestepped. Henri Becquerel was the first to discover radiation in 1896 while working in his Parisian lab. Along with his laboratory assistants Marie and Pierre Curie he conducted an experiment in a darkened room with a fragment of uranium exposed to photographic film. After a period of time it was noted that the film was fully exposed even though no visible light was apparently present in the room. But energy had escaped from the piece of uranium. The discovery of radiation and radioactive decay led to many other researchers entering the field, and Ernst Rutherford and Frederick Soddy found in 1902 that radioactive decay appeared to occur at a predictable rate for particular elements. It was soon proposed that this might actually provide a regular clock as a means of measuring the age of the earth, and Rutherford used his address at the Silliman Lecture at Yale in 1905 to challenge scientists to undertake this task. Rutherford believed that the discovery of radioactive elements provided an opportunity for the amount of time required by geologists and biologists to be increased sufficiently to account for the emergence of life.[5]

Bertram Boltwood made the first attempt at solving the Rutherford-Soddy challenge in 1907 with an estimation of 400 million to 2,200 million years for the age of the earth based on the known decay rate of uranium to lead. Arthur Holmes of Imperial College London further developed this estimation to 1,600 million years, again based on the decay rate of uranium to lead. He first published his results in 1913 and again in 1927 by which time his value was accepted. Through the 1930s and 1940s further studies by Holmes, Alfred Gerling, and others extended the age to 3,300 million years.[6]

The age of the earth became fixed in 1956 with the work of Claire Patterson of Caltech. His research decision was to focus on uranium to lead decay from measurements in iron meteorites that he thought were left over from the beginning of the solar system. This gave him a value of 4,600 million years, a value that is still accepted today by secular science.

Popular science programmes reinforce this air of authority with presenters speaking categorically when giving the earth's age, even though the evidence is based upon untestable assumptions and does not warrant such level of certainty.

Incidentally, while Claire Patterson fixed the age of the earth in the 1950s, new developments in genetics, with the discovery of DNA and the complexity of protein chemistry, revealed that even this length of time would not be sufficient to account for such complexity. This forms part of the modern intelligent design arguments involving calculations of a

---

[5] Baxter, *Revolutions in the Earth*, 2003, pp. 209-213
[6] Repcheck, J., *The Man Who Found Time: James Hutton*, 2003, pp. 199-207

universal probability bound. From this it can be shown that the complexity of many proteins exceeds the ability of the universe to produce them by accident, even if the whole universe were considered a pond of organic soup.

It is also interesting to note the close similarity between Claire Patterson's estimation of the age of the earth in terms of billions of years with the value given by de Maillet and the length of one day of Brahma. Recall that de Maillet argued in the eighteenth century for acceptance of the Hindu cosmology and timeframe for creation of two billion years, about a quarter of one day of Brahma. Today the 4.6 billion years of Claire Patterson is close to half a day Brahma. While this may appear only coincidental with modern scientific dating, it may be borne in mind that the concept of a 'path to enlightenment' is common to eastern religions, and was part of the western Enlightenment, especially as developed in Paris and Edinburgh. As shown both Lyell's gradualism and Darwin's evolutionary ideas can be traced back through Scottish and French Enlightenment philosophers who themselves had an interest in Greek and Eastern mysticism. There is a degree of commonality between the beliefs of Greek philosophers such as Aristotle and Plato and Hinduism through for instance the Greek translation of Berosus, and the work of Hesiod. Although similarity between the secular age of the earth may appear only coincidental with the cosmology of pagan religions, the notion that there is common ground is reinforced by the development of the idea of a cyclical universe involving periodic recreations.

The concept of *nature* is also ambiguous, especially in the writing of Spinoza and in others such as Hume, because it can be both applicable to an atheistic or a pantheistic cosmology. Bearing in mind the influence that pagan philosophies have had on the development of modern science it may be argued that the historical account given in the Judeo-Christian Scriptures has been undermined by another esoteric religious tradition. The idea that there is a battle between science verses Christianity is false; it may be argued instead that Christianity's opponent here is in effect a naturalistic religion that merely masquerades as objective science.

When we look to the philosophical foundations of science it can be seen that the main problem for accepting the age of the earth as sacrosanct is the methodology of science itself. Scientific investigations exist within competing research programmes, often seen as different paradigms. As such it is a danger to claim that one age for the universe must be accepted due to a particular dogma, however well regarded that dogma may be. Research programmes also move towards those that are deemed progressive in terms of their explanatory power, but old research programmes may be resurrected that were once thought dead. This is happening today to Flood geology as new discoveries arise.

The other problem that is often not recognised in attempts at dating the earth is that any derived values are dependent upon untestable assumptions. For instance, they depend on the regularity of radioactive decay rates through time. This claim is assumed and not measured, but further scientific research may find that under some circumstances such rates may change. Recent studies have suggested small changes in radioactive decay rates over a 33-day period, this related to the rotation of the inner core of the sun, and small annual changes as well due to the distance between earth and sun.[7] There is also the possibility that some other physical constants are not constant, but may have varied through time. Problems with the big bang theory and 'inflation' for instance led one group of scientists to discuss the possibility that the speed of light may have changed in the past. A varying speed of light would solve problems with the homogeneity of space, in relation to an expanding universe, that is unresolved in the current inflationary scenario.[8] Of course much more work needs to be done to determine whether so called 'constants' have varied through time, and the scale of possible changes. But it is not safe to assume that they have remained absolutely constant. The dogmatic insistence in favour of gradualism is in danger of being a 'science stopper' because it seeks to restrict discovery.

What is also unknown and unknowable is the amount of lead and uranium in iron-rich meteorites at the beginning of time. There is also the possibility that iron enriched meteorites may once have formed part of a planetary core that underwent the segregation of minerals due to the force of gravity. Estimates for the density and mass of the earth and other planets in orbit around the sun suggest that planetary cores are composed of iron. Thus the presence of iron rich asteroids in the asteroid belt may suggest that

---

[7] See for instance; Jenkins, J.H. et al. (2009). 'Perturbation of Nuclear Decay Rates During the Solar Flare of 13 December 2006'. *Astroparticle Physics* 31 (6): 407–411, and Jenkins, J.H. et al., (2009). 'Evidence of correlations between nuclear decay rates and earth–sun distance', *Astroparticle Physics* 32 (1): 42–46

[8] For instance: Magueijo, J., *Faster than the Speed of Light*, William Heinemann, London, 2003. or; Albrecht, A., Magueijo, J., *A 'Time Varying Speed of Light as a Solution to Cosmological Puzzles,' Phys. Rev. D* 59, 043516, 1999. or; Barrow, J.D., 'Cosmologies With Varying Light Speed,' *Phys. Rev. D*, 59, 043515, 1999. or; Moffat, J.W. 'Superluminary Universe: A Possible Solution to the Initial Value Problem in Cosmology,' *Int. J. Modern Physics, D*, Vol. 2, No. 3, pp. 351-365, 1993. or; Davies, P.C.W., Davis, T.M., Lineweaver, C.H., 'Black Holes Constrain Varying Constants,' *Nature,* 418 (6898), pp. 602-603, 2002.

some of these once formed the core of a planet that existed between Mars and Jupiter that subsequently broke up, although this is speculative.

The Flood geology of Steno, Burnet and Woodward had been successful in forcing acceptance of the organic origin of fossils against the plastic theory, and forced recognition of the sedimentary nature of many geological formations. However, the Flood model was unable to come up with a coherent theory as to how the waters could have covered the earth. Burnet had argued that the earth was cracked like an egg and that mountains were ruins of a former world, but others such as Newton and Hutton wanted to see in gradual geological processes an aspect of God's good design. So the catastrophic Flood model was rejected in the thinking of an increasingly deistic scientific community.

It wasn't really until the 1960s that fresh developments in geological science enabled a possible solution to the question of how the waters might have covered the earth. The theory of plate tectonics was developed in the 1960s, with the idea that continents were gradually moving around the planet. Continental drift was little known to earlier geologists. A German geologist, Alfred Wegener, described the theory of continental drift in 1915 with the publication of *The Origin of Continents and Oceans* involving an ancient continent called Pangaea. He noticed that the coastline of Africa and South America would fit together like a glove, and proposed that the continents were drifting around the globe at a rate of up to two and a half metres per year. However he had no mechanism to account for this drift, and his ideas were largely ignored at the time, although some scientists such as the Swiss Émile Argand supported Wegener as it helped to account for mountain building.

Through later decades scientists began to explore the deep ocean floor. This had been a major unknown for scientists previously. One attempt at investigation was led by a British expedition with the steam ship *HMS Challenger*. This team made many measurements between 1872 and 1876, but accurate readings were not possible until side scan sonar systems, and deep sea drilling were available from the 1960s onwards. The Deep Sea Drilling Project with the *Glomar Challenger* in the 1960s provided a great deal of evidence for the igneous nature of much of the base of the ocean floor, and this provided data for the development of plate tectonic theory.

The theory of plate tectonics seeks to explain how the ocean floor is created at the mid ocean ridges where the sea floor is diverging. Fresh molten magma rises from deep within the earth with the creation of a new sea floor surface. Because it is hotter and fresher than the old sea floor the line of upwelling forms mid-ocean ridges that are higher than the ocean basins. Ocean floor is also being destroyed at what are called subduction zones. Where plates are converging against one another, one plate is forced down under the other. The Pacific Plate for instance is being subducted under the South American Plate, and mountain ranges such as the Andes and Rockies are lifted up as a result of the ocean floor sliding under the continental plate. Where the oceanic plate is being forced downwards deep-ocean trenches will form.

What is of interest to modern Flood geologists is that plate tectonics is a convective process driven by heat within the earth. To understand plate tectonics it is necessary to view convection as a three dimensional process. Observe for instance how water circulates in a pan of boiling water and note what happens when the heat is turned up; convection increases. The question of interest then is whether the process of plate tectonics may be speeded up due to additional heat input. What would happen to the ocean floor and water if a catastrophic plate tectonic event were to occur? John Baumgardner is a leading theoretical geologist who has developed a well-regarded computer model (known as TERRA) to try and answer this question. When silicate rock is put under stress its viscosity increases by a factor of around one billion. Because the rock at the surface of the earth is colder than rocks in the depths, there is also a potential energy gradient. There is then a latent instability due to the stored energy on the earth that could lead to a rapid readjustment of the geological formations. If rock is forced into the earth, its stress and viscosity increases substantially and thus it could lead to a period of runaway subduction. But stored energy cannot be released unless there is additional energy put into the system to activate it. One possible source of activation energy would be the impact of a comet for instance, but there is sufficient stored energy in the rock itself to drive the process once started. A period of rapid plate tectonics would come to an end once the potential energy had been used up.

There are three factors that may lead to a global Flood as a result of catastrophic plate tectonics. But firstly, note that with the present earth covered by seventy percent ocean water, and with the average depth of the ocean around three kilometres, there is sufficient water to cover the present land surface to a depth of more than two kilometres. With that in mind we may note that the heat released in the process of rapid plate movements would be absorbed in the ocean water, and therefore the expansion of seawater would raise the sea level. Secondly the new ocean floor would also be hotter and be much higher than the old ocean floor, thus raising the sea level further. Thirdly with the new ocean floor considerably higher than the old ocean floor, then the colder areas of ocean floor and continental crusts would be drawn downwards due to isostatic compensation, again causing the sea level to rise relative to the land. There are some problems with this scenario, for instance it may be that too much heat is generated that might boil the oceans, although one possible solution to this is the impact of turbidity currents. These are sediment-rich mudslides that cover the sea floor and can travel at high speeds quickly locking in the heat from the hot upwelling rocks.

Chapter Summary

This chapter has looked at the way in which beliefs about the age of the earth developed following Darwin's book. Initially, following the thinking of Lord Kelvin, there was reluctance to accept Darwin's ideas because there was considered to be insufficient time for evolution to have occurred. Kelvin's reasoning was however criticised by Huxley. Later, the age of the earth developed to cover a period of 4.6 billion years by 1956, following the work of Claire Patterson and others. This, perhaps rather conveniently, corresponds reasonably closely with one day of Brahma, thus showing again the influence of pagan metaphysics in the development of secular an aspect of science. This was of course de Maillet's desire, and there is evidence that the influence of Greek and Hindu beliefs played their part in the development of geology in France and Scotland in the eighteenth century. This then fed into the thinking of nineteenth century protagonists. Although the Flood geologists had done important work in developing a biblical understanding through this period, there was a gap in knowledge concerning how the ocean waters might have covered the earth. And this was a real stumbling block. Subsequently however, with the development of plate tectonics, a new mechanism became available for Flood geologists; that is a more rapid version known as catastrophic plate tectonics. This may offer novel insights for the development of science. The assumptions that have gone into the secular age of the earth may also need revising given growing evidence that radiometric decay rates, and other constants of nature, may vary due to some processes that are as yet poorly understood.

15.

Conclusions

So this really brings us to the end of the discussion, and to the end of the book. This work has highlighted the influence of Greek and Hindu religious beliefs in the development of naturalistic science and acceptance of evolution and deep time. There is an ambiguity in the concept of nature, between atheism and pantheism, and this risks undermining the whole enterprise of science. This is because such ambiguity leads to relativism and to subjectivity in science. It is my belief that evolutionary naturalism is essentially a pagan narrative imposed upon science, as exampled for instance in the work of Erasmus Darwin in *Zoonomia*. Instead, science must begin with a belief in order, in objective truth, and that the universe is intelligible to the rational mind. For these reasons, true science is an enterprise that is dear to the heart of the Christian community. But it is not one that should hold the material world and spiritual realm entirely separate, as modern secular science seeks to do. However, this has not been the central theme of this book even though it needs stating.

It has been shown that beliefs in long periods of time and forms of evolution were present in ancient pagan cultures, with a focus in this study upon Greek mythology and philosophy, and Hinduism. But I have also noted the influence of Sumerian and Babylonian religions with their belief in the god of chaos, for instance as *Tiamat*.

Berosus, a priest of Baal, taking information from the Sumerian King List, extended the timeframe of the pre-Flood patriarchs for millennia, although it is noteworthy that Berosus remembered the biblical Flood. This work of Berosus was based upon the number units of six and sixty, while the biblical account involved creation over six days, followed by a seventh day of rest. The Hindu scriptures though extended hundred of thousands of years to millions and billions, although with influence from the Babylonian era seen in the continued use of a number system based upon the unit of six and sixty. Interestingly, this is also related to the number of seconds in a year. The reason for extended timeframes in antiquity may have been to do with poor translations and cultural misunderstandings, or for reasons of national prestige. Baal influence can also be seen in the way modern Hindus consider the cow to be a sacred animal for instance. Modern evolutionary ideas can also be seen in the pagan god of chaos *Tiamat*, and in the idea of an impersonal 'power of generation' at work in nature related to Hesiod's god of chaos.

Discussion has also touched upon the influence that Aristotle and Plato had upon the development of science, for instance in the failure of

Denslow, W.R., (1957-61) *10,000 Famous Freemasons*, 4 Vols., Missouri: Missouri Lodge of Research

Denton, M., (1982) *Evolution: A Theory in Crisis*, Bethesda, Maryland: Adler and Adler

Draper, J.W., (1881) *History of the Conflict Religion*, New York: D. Appleton and Co.

Duke of Argyll., (1887) 'A Reply: Science Falsely So Called,' *The Nineteenth Century* 21, May 1887, pp.771-774

Duke of Argyll., (1887) 'A Great Lesson,' *The Nineteenth Century* 22, Sept. 1887, pp.293-309

Fitzroy R., (1839) *Narrative of the Surveying Voyages of HMS Adventure and Beagle 1826-1836*, (A Very Few Remarks with Reference to the Deluge. Chapter XXVIII), Vol II, London: Henry Coburn

Forster, R and Marston, P. (1999) *Reason, Science and Faith*, Crowborough East Sussex: Monarch Books

Fuller, S., (2007*) Science and Religion? Intelligent Design and the problem of Evolution*, Polity Press

Galileo Galilei, (1968) [1615] 'Letter to Madame Christina of Lorraine, Grand Duchess of Tuscany: Concerning the Use of Biblical Quotations in Matters of Science,' translated by Drake, S., sourced from *Opere*, edited by Antonio Favaro, Giunti-Barbera, Firenze, 1968, Vol. V, pp. 309-348

Galling P and Mortenson, T., (2012) 'Augustine on the Days of Creation: A look at an alleged old-earth ally,' AiG Journal, Jan 18, 2012

Gould, S.J., (1975) 'Catastrophes and the Steady State earth,*' Natural History*, Vol. LXXX, No. 2, Feb 1975

Gould, S. J., (1984) 'Toward the vindication of punctuational change,' In: W. A. Berggren & J. A. Van Couvering (Eds.): *Catastrophes and earth History: The New Uniformitarianism*, Princeton N.J., Princeton University Press

Gribbin, J. & Gribbin, M., (2003) *FitzRoy*, Headline Book Publishing

Grinnell, G., (1976) 'A Probe Into The Origin of the 1832 Gestalt Shift in Geology,' *Kronos: A Journal of Interdisciplinary Synthesis* Glassboro, N. J: Kronos Press, Vol. 1(4), Winter 1976, pp. 68-76

Haeckel, E., (1874) 'Scientific Worthies: Thomas Henry Huxley,' *Nature*, February 1874, p.258

Harrison P., (2001) *The Bible, Protestantism and the Rise of Natural Science*, Cambridge: Cambridge University Press

Harrison, P., (2005) 'The Bible and the Emergence of Modern Science', *Christians in Science, Public Lecture*, Cambridge University, 24th May 2005

Hart Weed, J., (2007) 'De Genesi ad litteram and the Galileo Case,' invited article in *Go Figure! Essays on Figuration in Biblical Interpretation*, Ed. Stanley D. Walters., Eugene, OR: Wipf & Stock Publishers

Hasel, G., (1974) 'The Polemic Nature of the Genesis Cosmology', *Evangelical Quarterly*, Vol. 46, 1974, pp. 81-102.

Herbert, S., (1977) 'The Place of Man in the Development of Darwin's Theory of Transmutation: Part II, ' *Journal of the History of Biology*, Vol. 10, no. 2, 1977, p. 161

Himmelfarb, G., (1958) *Darwin and the Darwinian Revolution*, Chatto and Windus, p. 320

Hoskin, M. (1997) *Cambridge Illustrated History of Astronomy*, Cambridge: Cambridge University Press

Hume, D. (1854) 'Dialogues Concerning Natural Religion', in, *The philosophical works of David Hume [1711-1776]: Including all the essays, and exhibiting the more important alterations and corrections in the successive editions pub. by the author*, in Four Volumes, Volume II, p.411-540, Boston: Little, Brown and company; Edinburgh: Adam and Charles Black,

Hume, D. (1947) 'Dialogues Concerning Natural Religion,' in Kemp Smith, N., (ed.) *Dialogues Concerning Natural Religion*, 2nd ed. Indianapolis: Bobbs-Merrill Educational Publishing

Hutton, J., (1788) 'Theory of the Earth; or an Investigation of the Laws observable in the Composition, Dissolution, and Restoration of Land upon the Globe,' *Transactions of the Royal Society of Edinburgh*, Vol. I, Part II, 1788, pp.209-304

Huxley, T.H. (1854) On the Education Value of the Natural History Sciences, 1854

Huxley, T.H., (1859) 'Science and Religion,' *The Builder*, Vol. 17, Museum of Geology, January 1859, p. 35

Huxley, T.H., (1870) 'The Scientific Aspects of Positivism,' *Lay Sermons, Addresses and Reviews*, London, 1870, pp. 129-133

Huxley. T.H., (1870) 'Life in the Deep Sea,' *Nature*, 2nd July 1870, p. 187

Huxley. T.H., (1875) 'Notes from the Challenger,' *Nature*, 12, Aug., 1875, pp.315-316

Huxley. T.H., (1879) 'Report to BAAS,' *Nature*, 28th August

Huxley, T.H., (1893) 'An Episcopal Trilogy' (*The Nineteenth Century*, 22, November 1887, pp. 625-40), in *Collected Essays*, Vol. V, London, pp.126-159

Huxley, T.H., 'Geological Reform' (1869 Lay Sermons), in *Collected Essays*, Vol. VIII, London, 1893, pp. 305-399

Huxley, T.H., (1893), 'Scientific Education: Notes of an After-dinner Speech' (given in 1869 at the Liverpool Philomathic Society), in *Collected Essays*, Vol. III, London

Irvine, W., (1956) *Apes, Angels and Victorians*, London: Weidenfeld and Nicolson, 1956

Jaki, S.L., (1974) *Science and Creation*, Edinburgh: Scottish Academic Press

Jenkins, J.H. et al. (2009) 'Perturbation of Nuclear Decay Rates During the Solar Flare of 13 December 2006'. *Astro-Particle Physics,* 31 (6): 407–411

Jenkins, J.H. et al., (2009) 'Evidence of correlations between nuclear decay rates and earth–sun distance', *Astro-Particle Physics,* 32 (1): pp. 42–46

Jensen, J.V., (1970) 'The X-Club,' *British Journal for the History of Science*, pp. 59, 63-72, 179.

Jones, F.A., (1912) [1909] *The Dates of Genesis*, Kingsgate Press

Lany, J.P., and Berthault G., (1993) 'Experiments on stratification of heterogeneous sand mixtures', *Bulletin of the Geological Society, France*, pp. 164-5 and 649-660

Lennox, J., 2012, *Seven Days that Divide the World*, Grand Rapids: Zondervan

Lucas, E., (2004) 'Science and the Bible, Are they Compatible?,' *Christians in Science - St Edmunds College Public Lecture*, 13th May 2004

Lucas, E., (2005), 'Science and the Bible, Are they Compatible?,' *Science and Christian Belief*, Vol. 17, pp. 137-154

Keynes, R., (2002) *Fossils, Finches and Fuegians: Charles Darwin's Adventures and Discoveries on the Beagle 1832-1836*, London; Harper Collins

Kirby, W., (1835) *The Bridgewater Treatise, Treatise VII*, Vol. I, London: William Pickering

Kirwan, R., (1799) *Geological Essays*

Lambert, W.G, and Walcot, P., (1965) 'A New Babylonian Theogony and Hesiod', *Kadmos* 4 (1), pp. 64-72

Lindberg, D.C., & Numbers R.L., (1986) (eds.) 'Introduction,' in *God & Nature: Historical Essays on the Encounter between Christianity and Science*, Berkeley: University of California Press, 1986, pp. 2-3

Linden, S.J. (2003) *The Alchemy Reader: From Hermes Trismegistus to Isaac Newton*, New York: Cambridge University Press

Lopez, R.E., (1998) 'The Antidiluvian Patriarchs and the Sumerian King List', *CEN Tech. J.* 12 (3), pp. 347-357

Lucas, E., (2004) 'Science and the Bible, Are they Compatible?,' *Christians in Science - St Edmunds College Public Lecture*, 13th May 2004

Lyell, C., (1834) *Principles of Geology*, Vols. I-IV, London, John Murray

Lyell, K, (1881) *Life, Letters and Journals of Sir. Charles Lyell*, (ed.) London: John Murray

Magueijo, J., (2003) *Faster than the Speed of Light*, William Heinemann, London

Malthus, T.R., (1826) *Principles of Population*, 6th edition, Ward Lock & Co.

Mantell, G., (1831) 'The Geological Age of Reptiles,' *New Philosophical Journal*, Vol. XI, Edinburgh, Apr-Oct, pp. 181-185

Marston, P., (2007) 'Johannes Kepler,' *CIS Magazine (Online)*, No.4, Autumn 2007

Maunder, E.W., (1909) 'Book Review: The Dates of Genesis,' *The Observatory*, Vol. 32, pp. 390-393

McCalla, A., (2006) *The Creationist Debate*, London: Continuum

McGrath, A.E., (2002) *Christian Theology: An Introduction*, 3rd edition, Blackwell

Moffat, J.W., (1993) 'Superluminary Universe: A Possible Solution to the Initial Value Problem in Cosmology,' *Int. J. Modern Physics D*, Vol. 2, No. 3, pp. 351-365

Mortenson, T., (1997) British scriptural geologists in the first half of the nineteenth century- part 1: Historical Setting, TJ (Journal of Creation), 11(2), 1997, pp. 221-252

Mortenson, T., (2004) *The Great Turning Point*, Grand Rapids: Master Books

Penn, G., (1825) *Comparative Estimate of the Mineral and Mosaical Geology*, Vol. I, pp. 150-152

Plantinga, M., 'Methodological Naturalism?' in Pennock, R.T. (ed.) *Intelligent Design Creationism and its Critics: Philosophical, Theological and scientific perspectives*, Cambridge Massachusetts, The MIT Press, 2001, p.339

Polanyi, M., (1946) *Science, Faith and Society*, London: Oxford University Press

Popper, K., (1957) *The Poverty of Historicism*, London: Routledge and Kegan Paul

Pulman, G.P.R., (1975) [1875] *The Book of the Axe*, 4th ed. Kingsmead Reprints: Bath, 1975.

Repcheck, J., (2003) *The Man who Found Time: James Hutton*, New York - London: Simon & Schuster

Ritland, R., (1981) 'Historical Development Of The Current Understanding Of The Geologic Column: PART I,' *Origins*, 8(2), Geoscience Research Institute, 1981, pp. 59-76

Roberts, M., (2004) 'Intelligent Design; some Geological, Historical, and Theological Questions,' in Dembski, M and Ruse, M *Debating Design*, Cambridge University Press, Cambridge

Rohl, D., (1998) *Legend: The Genesis of Civilisation*, Century

Ross, H., 1994, *Creation and Time: A Biblical and Scientific Perspective on the Creation-Date Controversy.* Colorado Springs, CO: Navpress

Rudwick, M.J.S., (1972) *The Meaning of Fossils*, Chicago: Chicago University Press

Rudwick, M.J.S., (1985) *The Meaning of Fossils: Episodes in the History of Palaeontology*, Chicago: University of Chicago Press

Rudwick, M.J.S., (2008) *World Before Adam: The Reconstruction of Geo-History in the Age of Reform*, Chicago: University of Chicago Press

Schaff, P., (1885-1890) Anti-Nicene Fathers, (10 vols. first published 1885), and Nicene and Post Nicene Fathers, (28 Vols. Series I and II, 1886-1890) available online at www.ccel.org

Schneer, C.J., (1967) *Towards a History of Geology*, Bourdier, F., Boston Mass: MIT Press, 1967

Sedgwick, A. (1969) [1834] *Discourses on Studies of the University* The Victorian Library, Leicester University Press

Sedley, D., (2007) *Creationism and its Critics in Antiquity*, LA: University of California Press

Sharratt, M., (1994) *Galileo: Decisive Innovator*, Cambridge: Cambridge University Press

Sibley, A, 'FitzRoy, Captain of the Beagle, Fierce Critic of Darwinism,' *Impact*, 389, ICR, November 2005.

Sibley, A, 'Bathybius and a Reign of Terror,' *Journal of Creation*, 23(1), CMI, pp. 123–127, April 2009.

Sibley, A., 'Creationism and millennialism among the Church Fathers', *Journal of Creation*, 26(3), CMI, pp. 95-100, December 2012

Snobelen, S., (2008) 'This most beautiful system': Isaac Newton and the Design Argument', in, *God, Nature and Design Conference*, Ian Ramsey Centre, Oxford, 10-13 July 2008

Spokes, S., (1927) *Gideon Algernon Mantell, LLD, FRCS, FRS, Surgeon and Geologist*, London: John Bale and Sons and Danielson

Steno, N, (1916) (English Translation) Winter, J.G., *The Prodromus of Nicholaus Steno's Dissertation Concerning a Solid Body Enclosed by Process of Nature Within a Solid*, New York & London: Macmillan Co. Ltd.

Stott, R, *Darwin's Ghosts: In Search of the First Evolutionists*, London: Bloomsbury Publ.

Strauss, L., (1952) *Persecution and the art of writing*, Chicago: Chicago University Press

The Eye (1999) 'Erasmus Darwin Centre opens in Lichfield,' *Freemasonry Today*, Issue 9, Summer 1999

Winchester, S., (2001) *The Map that Changed the World: William Smith and the Birth of Modern Geology*, New York: Harper Collins

Whitcomb J., and Morris, H, (1961) *The Genesis Flood*, Baker Book House

White, A.D., (1917) *Autobiography of Andrew Dixon White*, Vol. II, New York: The Century Co.

White, A.D., (1896) *A History of the Warfare of Science with Theology in Christendom*, (Chapter V, Part IV) New York: D. Appleton and Co.

White, A.D., (1960) [1896] *A History of the Warfare of Science with Theology*, Vol. 1, Reprinted New York: Dover Publ.

Whitehurst, J., (1778) *An Inquiry into the Original State and Formation of the earth*, London, (2nd ed. 1786, 3rd ed. 1792)

Zittel, K., (1901) *History of Geology and Palaeontology*, London: Walter Scott

Zuiddam, B., (2010) 'Augustine: young earth creationist,' *Journal of Creation*, 24(1), pp. 5-6

Index

CPSIA information can be obtained
at www.ICGtesting.com
Printed in the USA
BVHW042120030319
541675BV00005B/28/P

9 780956 214614